COMPARATIVE SKELETAL ANATOMY

COMPARATIVE SKELETAL ANATOMY

A PHOTOGRAPHIC ATLAS FOR MEDICAL EXAMINERS, CORONERS, FORENSIC ANTHROPOLOGISTS, AND ARCHAEOLOGISTS

By

BRADLEY J. ADAMS, PhD
Office of Chief Medical Examiner,
New York, NY

PAMELA J. CRABTREE, PhD
Department of Anthropology, New York University,
New York, NY

Photographs by

GINA SANTUCCI

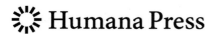 Humana Press

Authors
Bradley J. Adams, PhD
Office of Chief Medical Examiner
520 1st Avenue
New York, NY 10016
badams@ocme.nyc.gov

Pamela J. Crabtree, PhD
Department Anthropology
New York University
25 Waverly Place
New York, NY 10003
pc4@nyu.edu

ISBN: 978-1-58829-844-7   e-ISBN: 978-1-59745-132-1

Library of Congress Control Number: 2008921061

While the advice and information in this book are believed to be true and accurate at the date of going to press, neither
the authors nor the editors nor the publisher can accept any legal responsibility for any errors or omissions that may be
made. The publisher makes no warranty, express or implied, with respect to the material contained herein.

Cover illustration: bear skull showing upper and lower dentition (*see* discussion in Chapter 4).

Printed on acid-free paper

9 8 7 6 5 4 3 2 1

springer.com

PREFACE

Bones are frequently encountered in both archaeological and forensic contexts. In either situation it is critical that human remains are differentiated from non-human remains. In the realm of forensic investigations, this is usually the final determination. In the archaeological context, greater precision in identification may be warranted in order to draw conclusions about ancient diets, animal husbandry and hunting practices, and environmental reconstructions. This photographic atlas is designed to assist the archaeologist or forensic scientist (primarily zooarchaeologists and forensic anthropologists) in the recognition of various species that are commonly encountered in both contexts. Obviously the ability to differentiate between the bones of various species (let alone simply human vs non-human bones) is dependent upon the training of the analyst, but good reference material is also essential. While there are books dedicated to human osteology and books that focus on animal osteology, there is really nothing that brings the two together. It is our intent to fill this void with the compilation of photographs presented in this atlas. Greater attention is given to the postcranial remains, which are presented in standard anatomical orientations. In addition, "non-traditional" photographs of the various non-human species are also included in an attempt to bring together both anatomical and artistic images.

For this atlas, the large, non-human mammals include: horse (*Equus caballus*), cow (*Bos taurus*), black bear (*Ursus americanus*), white-tail deer (*Odocoileus virginianus*), pig (*Sus scrofa*), goat (*Capra hircus*), sheep (*Ovis aries*), and dog (*Canis familiaris*). All of these are compared to a modern adult male human skeleton.

The smaller non-human animals include: raccoon (*Procyon lotor*), opossum (*Didelphis virginiana*), cat (*Felis catus*), rabbit (*Oryctolagus cuniculus* and *Sylvilagus floridanus*), turkey (*Meleagris gallopavo*), duck (*Anas platyrhynchos*), chicken (*Gallus gallus*), rat (*Rattus norvegicus*), red fox (*Vulpes vulpes*), and snapping turtle (*Chelydra serpentina*). All of these are compared to a modern newborn human skeleton.

The first part of this book consists of a brief introduction followed by detailed black and white photographs of the key postcranial elements from the animals listed above. In order to show size and shape variations between the human and the non-human species selected for this atlas, scaled skeletal elements are pictured side-by-side. For example, a cow humerus and a human humerus are placed side-by-side in order for the reader to observe how they differ. Anterior (i.e., front or cranial in animals) and posterior (i.e., back or caudal in animals) views of each bone are presented. In some cases, medial or lateral views are also included.

The second part of the book consists of an overview of common butchering techniques used in traditional and commercial meat processing. This is followed by photographs of representative butchered bones. We have included a range of different butchery marks, including both prehistoric cut marks made with stone tools and historic cut marks made with cleavers and saws. We have also included examples of sawn human bones from a forensic case associated with intentional body dismemberment. Since bone was a common raw material throughout antiquity and up until the early 20th century, we have also illustrated a number of examples of worked bone artifacts.

Overall, we hope that this book will fill a void in the forensic science and archaeological literature, presenting comparisons between human and non-human bones that are useful to the archaeologist and forensic scientist. It is our goal that this book is frequently consulted as a laboratory and field reference guide...one that gets worn and discolored over the years from continued use and not a book that sits idle on a book shelf.

Bradley J. Adams
Pamela J. Crabtree

CONTENTS

Preface v

About the Author ix

  1 Introduction 1

  2 Human vs Horse 9

  3 Human vs Cow 29

  4 Human vs Bear 45

  5 Human vs Deer 75

  6 Human vs Pig 97

  7 Human vs Goat 117

  8 Human vs Sheep 133

  9 Human vs Dog..153

10 Human vs Raccoon 177

11 Human vs Opossum 195

12 Human vs Cat 217

13 Human vs Rabbit..235

14 Human vs Turkey 251

15 Human vs Duck..269

16 Human vs Chicken 289

17 Miscellaneous 307

18 Traces of Butchery and Bone Working..323

19 References 347

ABOUT THE AUTHORS

Bradley J. Adams

Bradley J. Adams received his BA from the University of Kansas and his MA and PhD degrees from the University of Tennessee. He is currently the Director of the Forensic Anthropology Unit for the Office of Chief Medical Examiner (OCME) in New York City. He is also affiliated with numerous universities in the New York City area. In his present position with the OCME, Dr. Adams and his team are responsible for all forensic anthropology casework in the five boroughs of New York City (Manhattan, Brooklyn, Queens, the Bronx, and Staten Island). Prior to accepting the position in New York City, Dr. Adams was a forensic anthropologist and laboratory manager at the Central Identification Laboratory in Honolulu, Hawaii.

Pamela J. Crabtree

Pamela J. Crabtree is an associate professor of anthropology at New York University, where she has taught since 1990. Her area of specialization is zooarchaeology, and she has analyzed a wide variety of faunal collections from late prehistoric and early medieval Europe, the Middle East, and historic North America. Dr. Crabtree is co-author of *Exploring Prehistory: How Archaeology Reveals Our Past* (2006) and she is co-editor of *Ancient Europe: Encyclopedia of the Barbarian World 8000 BC–AD 1000*. She is currently a member of the archaeological team that is surveying the Iron Age site of Dún Ailinne in Ireland.

1 Introduction

Regardless of the context (forensic or archaeological), the correct identification of human and non-human remains is a very serious issue in osteological analyses. While the difference between various species is often very striking, it can also be quite subtle (Figure 1-01). Case studies and text books have highlighted similarities between some species, for example the hand and foot bones (metacarpals and metatarsals) of the human hand and the bear paw in the forensic realm (Byers 2005; Owsley and Mann 1990; Stewart 1979; Ubelaker 1989). These comparisons between the human and bear are also presented in Chapter 4 of this book. Sometimes the morphological similarity between species is quite unusual and counterintuitive. For example, there is a remarkable correspondence between an adult human clavicle and an adult alligator femur (Figure 1-02).

The goal of this book is to create a comprehensive photographic guide for use by experienced archaeologists and forensic scientists to distinguish human remains from a range of common animal species. The atlas illustrates the larger mammal species in comparison to adult human bones, while the smaller mammal, bird, and reptile species are compared to an infant human skeleton. We have chosen to photograph the Old World domesticates—cattle (*Bos taurus*) , sheep (*Ovis aries*) , goat (*Capra hircus*), horse (*Equus caballus*), and pig (*Sus scrofa*)—since these animals are frequently found on historic archaeological sites in North America, and they are commonly recovered from Neolithic and later sites in the Eastern Hemisphere. Furthermore, they are also common in modern contexts and could easily end up being submitted as a forensic case.

The atlas includes three domestic bird species; two of them, chicken (*Gallus gallus*) and duck (*Anas platyrhynchos*), were initially domesticated in the Eastern Hemisphere, while the third, turkey (*Meleagris gallopavo*), was first domesticated by Native Americans. We have also chosen to illustrate a range of North American wild mammals, including many that were frequently hunted by Native Americans in pre-Columbian and colonial times. These include black bear (*Ursus americanus*), white-tail deer (*Odocoileus virginianus*), raccoon (*Procyon lotor*), and opossum (*Didelphis virginiana*). We have also included two species of rabbits. The smaller rabbit is the native wild rabbit or cotton-tail (*Sylvilagus floridanus*), while the larger rabbit is a domestic rabbit (*Oryctolagus cuniculus*) which is originally of European origin. Commensal species are frequently found in historic-period archaeological sites, and we have illustrated two of the most common, dog (*Canis familiaris*) and cat (*Felis catus*). We have also included a chapter of miscellaneous photographs (Chapter 17). In this chapter various views are presented of infant and adult human skeletons, selected comparisons between human and red fox (*Vulpes vulpes*), bobcat (*Felis rufus*), rat (*Rattus norvegicus*), and snapping turtle (*Chelydra serpentina*). The snapping turtle is the only reptile that is included as many of the bones are distinctive is shape and they are commonly recovered from North American archaeological sites.

1

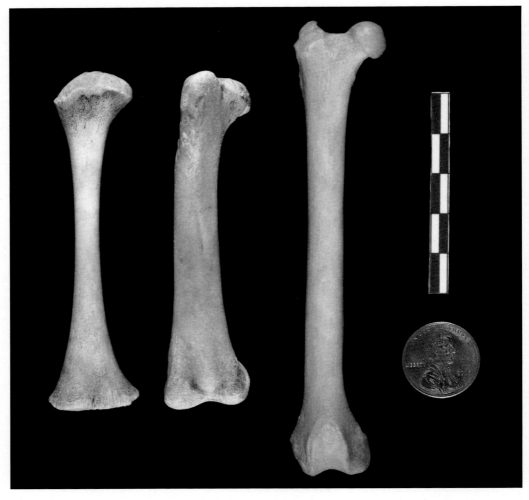

Fig. 1-01. Comparison from left to right of infant human, adult chicken, and adult cat right femora (anterior views).

Most archaeological faunal remains are the leftovers from prehistoric and historic meals. Many animal bones show traces of butchery that reveal the ways in which the carcass was dismembered. Furthermore, it is not unusual for food refuse to be mistaken for human remains and end up in the medical examiner or coroner system. In this atlas we have illustrated a range of different butchery marks and techniques (Chapter 18), including both prehistoric cut marks made with stone tools and historic cut marks made with cleavers and saws. We have also included examples of sawn and butchered faunal bones and have included schematic diagrams of modern, commercial butchery patterns. Since bone was a common raw material throughout antiquity and up until the early 20th century, we have also illustrated a number of examples of worked bone artifacts. Finally, knife cuts and saw marks in bone are not unique to non-human remains. There are numerous cases each year of intentional body mutilation using knifes and/or saws. In cases of human dismemberment (usually implying sawing through bones) or disarticulation (usually implying separation between joints) it is quite possible that a badly decomposed or skeletonized human body portion may

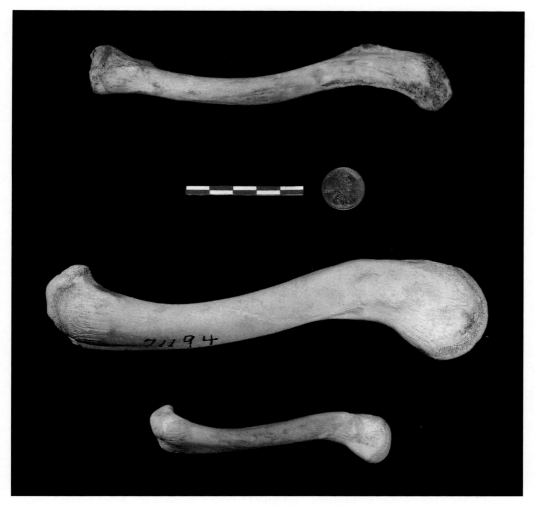

Fig. 1-02. Comparison of an adult human clavicle with alligator and crocodile femora; note the similar morphology between the human and nonhuman elements. Top is a left human clavicle, middle is a right *Crocodylus acutus* femur, bottom is a right *Alligator mississippiensis* femur.

appear non-human to the untrained eye. A forensic example of postmortem human dismemberment is also presented in Chapter 18 to show the similarity of tool mark evidence.

The ability to differentiate between complete or fragmentary human and non-human bones is dependent on the training of the analyst and the available reference and/or comparative material. It is truly a skill that requires years of training and experience and is not something that can be gleaned entirely from books. There is no substitute for coursework and training in osteology with actual skeletal material in order to appreciate the range of variation within all animal species. An experienced osteologist should always be consulted for confirmation of element type and species if there is any doubt.

ARCHAEOLOGICAL CONTEXT

Animal bones have played critical roles in archaeological interpretation for more than one hundred and fifty years of scientific endeavors. The discovery of the bones of extinct animals in association with simple chipped stone tools in sites in France and Britain helped to establish the antiquity of the human presence in Europe and to overthrow the traditional 6000-year biblical chronology for human life on earth. Faunal remains have also played a crucial role in the reconstruction of early human subsistence practices, in the study of animal domestication in both the Eastern Hemisphere and the Americas, and in the analysis of the ways in which historic cities were provisioned with food. Large numbers of animal bones are often recovered from archaeological sites, and these bones can be used to study past hunting practices, animal husbandry patterns, and diet. In order to use animal bones in archaeological interpretation, zooarchaeologists (archaeologists who specialize in the study of faunal remains) must be able to identify the bones, determine sex and age at death when possible, and examine the bones for evidence of butchery marks and traces of bone working.

While archaeologists expect to find human remains in cemeteries, human bones are often found in other contexts. For example, two adult human burials and the remains of several infants were unexpectedly recovered from the habitation area of the early Anglo-Saxon village site of West Stow in eastern England (West 1985: 58-59). This was the case even though the settlement site was associated with a nearby contemporary cemetery. In another example, at the late Neolithic site of Hougang near Anyang in China, burials of infants in pits or urns were associated with house construction activities (Chang 1986: 270). In short, zooarchaeologists and physical anthropologists must be able to confidently identify both animal bones and human remains in order to accurately interpret past cultures.

The first step in the analysis of animal bones recovered from archaeological sites is the careful identification of both body part and animal species. Precise identification requires a good comparative collection of modern specimens whose species, sex, and age are well-documented. However, a comparative collection must be supplemented by identification guides and atlases that can help the researcher distinguish between different species. Most zooarchaeological identification guides focus solely on non-human species, (e.g., Brown and Gustafson 1979; Cornwall 1956; Gilbert 1990; Gilbert, et al. 1981; Olsen 1964, 1968) even though human remains are commonly found in archaeological sites. One exception to this is Schmid (1972) who does illustrate human bones, but there is no comparison with subadult human bones.

FORENSIC CONTEXT

It is equally important for forensic scientists working with human skeletal remains to be able to differentiate between human and non-human bones. In the modern forensic context, it is quite common for non-human bones to be mistaken for human remains and end up in the medical examiner or coroner system. It is of obvious importance that they are correctly identified as such, or the consequences could be substantial. It is usually the role of a forensic anthropologist to make this assessment of "human vs. non-human" and generate the appropriate report. In most forensic scenarios, once a determination of non-human is made it is seldom of investigative significance to correctly identify the species. There are numerous skeletal anatomy books dedicated to human osteology (e.g., Bass 2005; Brothwell 1981; Scheuer and Black 2000; Steele and Bramblett 1988; White 2000; White and Folkens 2005). Some guides and textbooks on human osteology and forensic anthropology do include sections on differentiating between human and non-human remains (e.g., Bass 2005; Byers 2005; Ubelaker 1989) but these are more cursory discussions.

When attempting to differentiate between human and non-human skeletal remains, fragmentation only compounds the problem. If fragmentation is so extreme that gross identification of human versus non-human bone is not possible, microscopic (i.e., histological) techniques can be employed (e.g., Mulhern and Ubelaker 2001). Under magnification, the shape of the bone cells may be indicative of non-human bone, but this technique is not "fool proof" as some non-human animals (e.g., large dogs, bovines, and non-human primates) are nearly identical to humans microscopically. Our atlas will only focus on the gross assessment of bones.

BOOK TERMINOLOGY AND ORGANIZATION

In constructing this atlas, we have chosen to illustrate examples of both adult and juvenile animal bones, in addition to adult and infant human skeletons. Other guides to the identification of birds and mammals from archaeological sites illustrate only adult bones. However, many animal bones recovered from archaeological sites and within the forensic context are the remains of juvenile animals. Farmers who keep cattle for milk, for example, often slaughter excess male calves during their first year of life. In a meat-oriented economy, farmers frequently choose to slaughter adolescent animals, since these animals are nearly full-grown, and continuing to feed animals beyond adolescence results in only limited increases in meat output. We have included illustrations of both adult and juvenile pigs, and we have illustrated both an adult sheep and an immature goat. We have also photographed examples of immature chickens, since most chickens consumed today are quite young.

In general, the animals in this atlas are presented in the order of their size, progressing from largest to smallest. The corresponding human and non-human elements are presented alongside each other in order to fully appreciate the variation in size and shape between them. In order to add a scaled perspective, a metric ruler (centimeters) is present in each photograph along with a U.S. penny.

Bipedalism, upright walking on two legs, is one of the most important developments in all of human evolution. However, as a result of bipedalism, many human bones are oriented in somewhat different ways than comparable bones are in other mammals. In addition, the directional terms used to describe parts of the body differ somewhat between humans and other mammals (Figures 1-03 and 1-04). For example, in human osteology the term *anterior* is used to describe the front portion of a bone, while in quadrupeds the term *cranial* is used. Similarly, the back portion of the femur is described as *posterior* in humans, but it is described as *caudal* in other mammals. Different terms are also used for the lower portions of mammal limbs. For example, the surface of the forelimb (distal to the radius and ulna) that faces the ground is described as *palmar* (or *volar*), while the comparable surface in the hindlimb is described as *plantar*. The opposite surfaces of the bone are described as *dorsal*. The terms *proximal, distal, medial,* and *lateral* are used to describe surfaces in both human and non-human bones. For humans, we have used the directional terms as described in Bass (2005). For other mammals, they have used the terms as defined in Evans and de Lahunta (1980) and Getty (1975). In describing bird bones, we have followed the terminology used by Cohen and Sergeantson (1996).

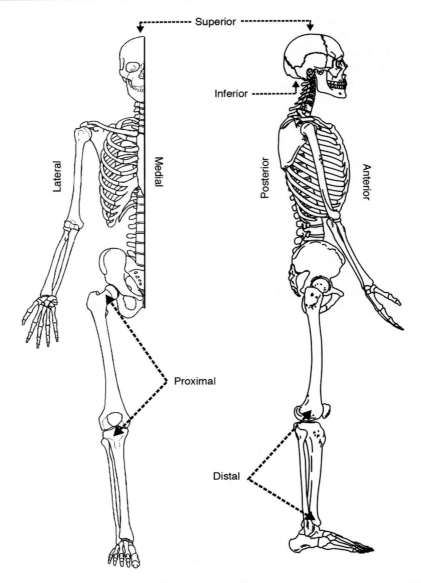

Fig. 1-03. Schematic diagram of human skeleton in standard anatomical position (i.e., standing with arms at the side and palms forward so that no bones are crossing) labeled with anatomical terminology.

BACKGROUND OF THE SPECIMENS INCLUDED IN THIS BOOK

Most of the non-human skeletons that are illustrated in this atlas come from the collections of the zooarchaeology laboratory in the Anthropology Department of New York University. The bear skeleton was borrowed from the Department of Mammology of the American Museum of Natural History. Most of the horse bones that are illustrated here are from a horse skeleton that was borrowed from the Museum Applied Science Center for Archaeology (MASCA) at the University of Pennsylvania Museum.

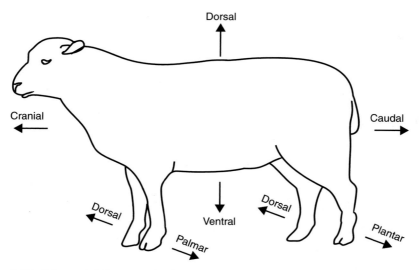

Fig. 1-04. Schematic Diagram of animal skeleton labeled with anatomical terminology.

The raccoon skeleton was borrowed from Susan Antøn. The alligator and crocodile femora were provided by the Herpetology Department at the American Museum of Natural History and were photographed by Ilana Solomon and Tam Nguyen. The original photograph of the turkey skull was provided courtesy of the National Wild Turkey Federation, while Gina Santucci performed the artistic modifications to the photograph. Seth Brewington provided the photograph of the antler comb from Iceland. The horse metacarpus and metatarsus were borrowed from the Zooarchaeology Laboratory in the Anthropology Department at Hunter College. Jeannette Fridie was a great help with many facets of this book. The human remains are from unidentified individuals that were analyzed at the Office of Chief Medical Examiner in New York City. We are grateful to everyone who loaned us specimens and assisted in this project.

2   Human vs Horse

Fig. 2-00. A lateral view of the horse's cranium. The horse's dental formula is 3/3.0-1/0-1.3/3.3/3. Canines are usually seen only in males.

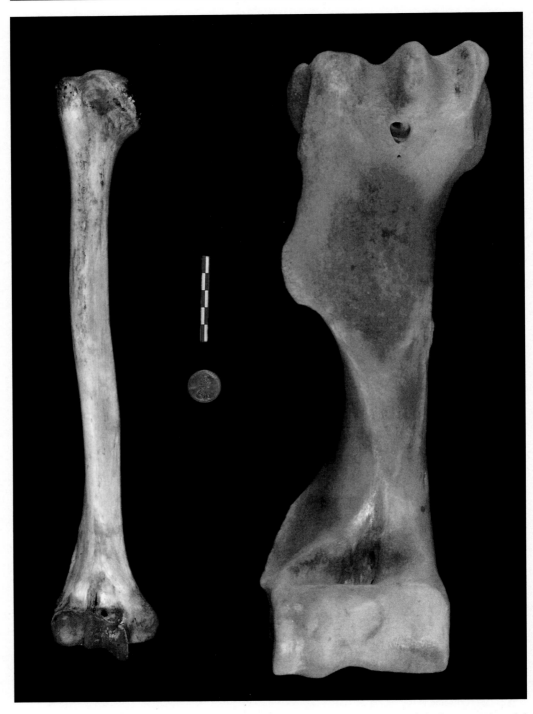

Fig. 2-01. A human right humerus (anterior view) is compared to a horse's right humerus (cranial view). The shaft of the horse's humerus has a large deltoid tuberosity. The proximal end of the horse's humerus includes an intermediate tubercle, which is not seen on the human humerus.

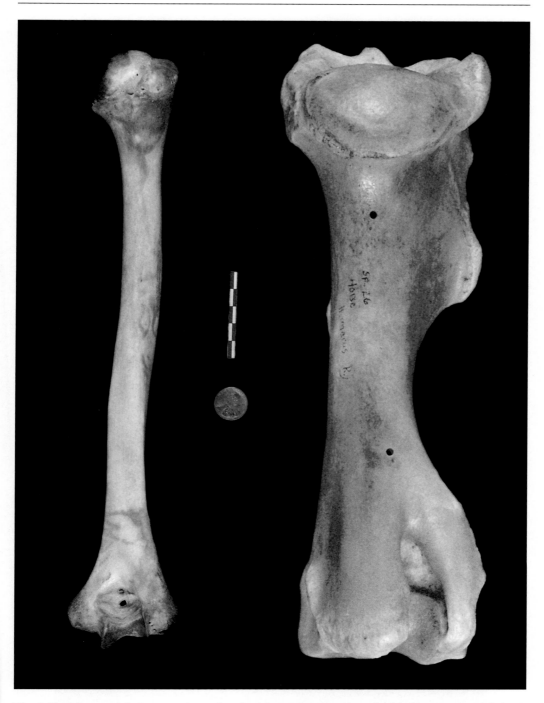

Fig. 2-02. A human right humerus (posterior view) is compared to a horse's right humerus (caudal view).

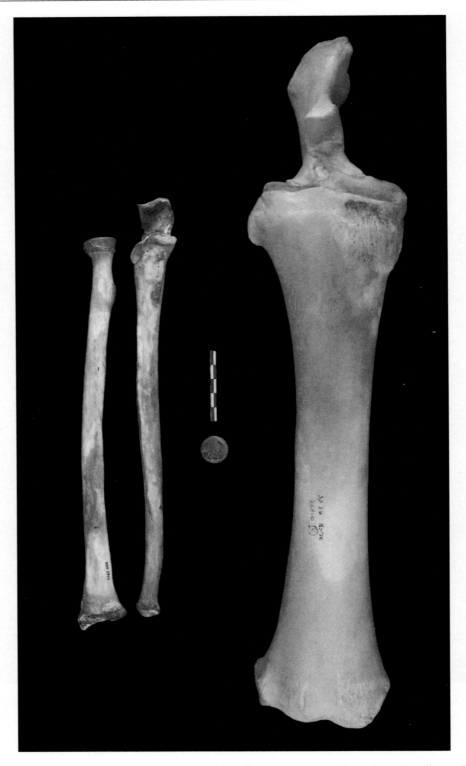

Fig. 2-03. A human right radius and ulna (anterior views) are compared to a horse's radius and ulna (cranial view). Note the large olecranon process on the horse's ulna.

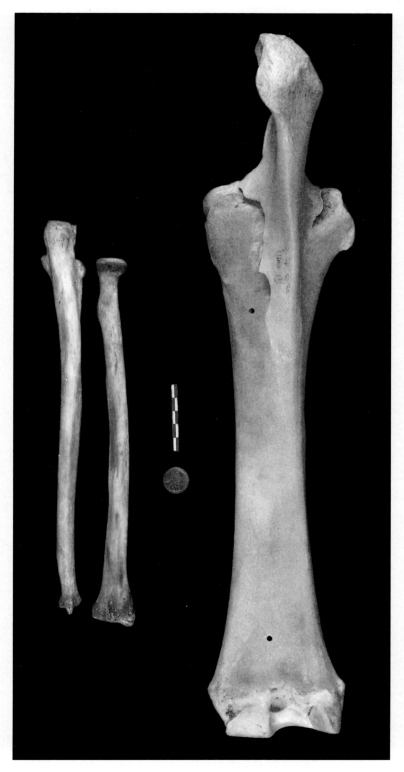

Fig. 2-04. A human right radius and ulna (posterior views) are compared to a horse's radius and ulna (caudal view). Note that the horse's ulna tapers to a point about two-thirds of the way down the shaft of the radius.

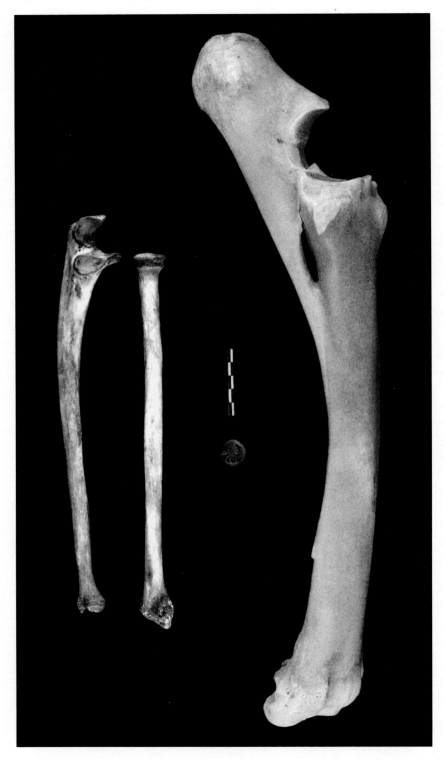

Fig. 2-05. A human right radius and ulna (lateral views) are compared to a horse's radius and ulna (lateral view). The horse's ulna is partially fused to the radius in adults.

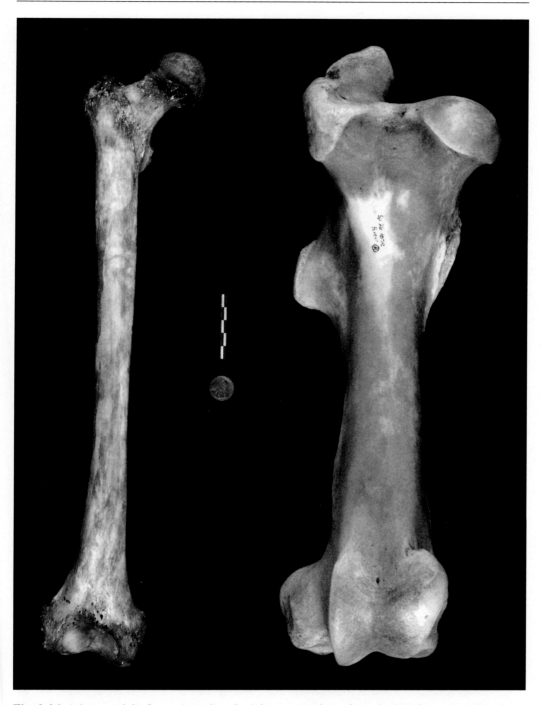

Fig. 2-06. A human right femur (anterior view) is compared to a horse's right femur (cranial view). The horse's femur shows a well developed third trochanter.

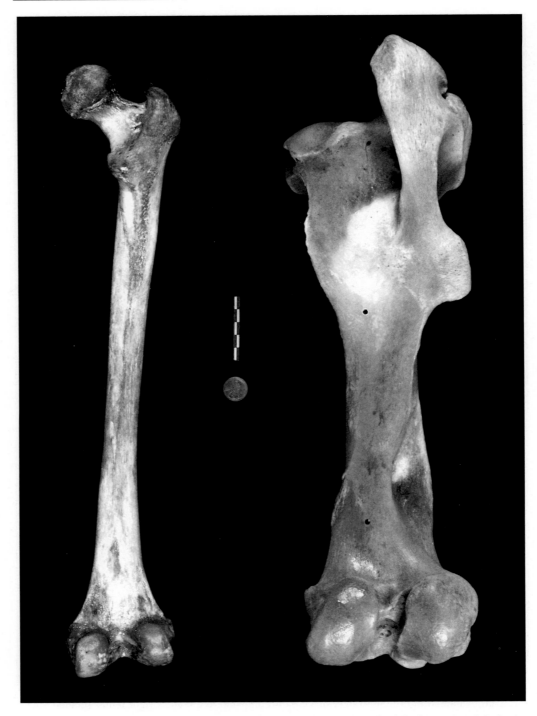

Fig. 2-07. A human right femur (posterior view) compared to a horse's right femur (caudal view).

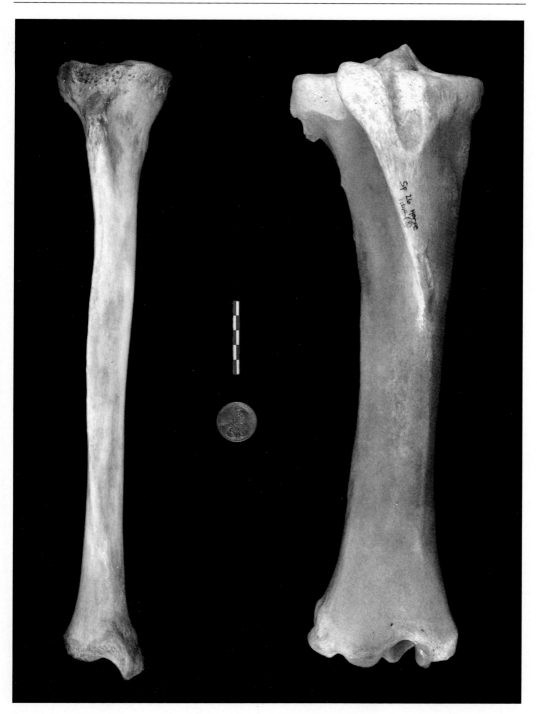

Fig. 2-08. A human right tibia (anterior view) is compared to a horse's right tibia (cranial view). The horse distal tibia includes both a medial and a lateral malleolus. The lateral malleolus is the evolutionary remnant of the distal fibula.

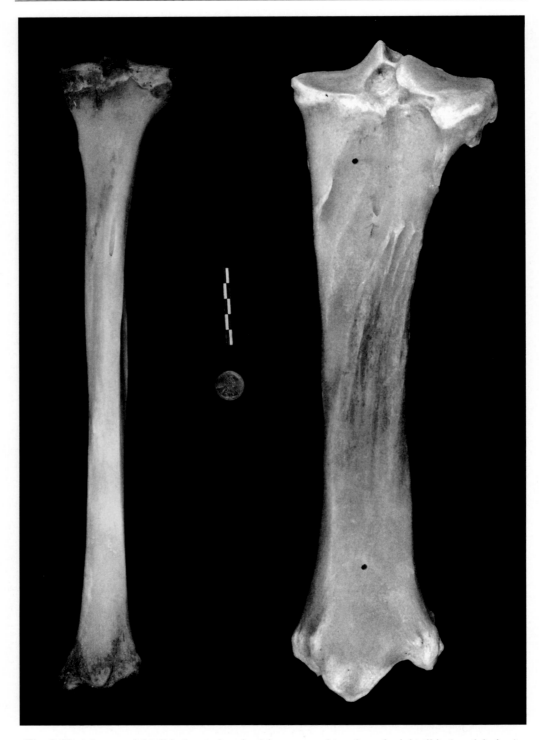

Fig. 2-09. A human right tibia (posterior view) is compared to a horse's right tibia (caudal view).

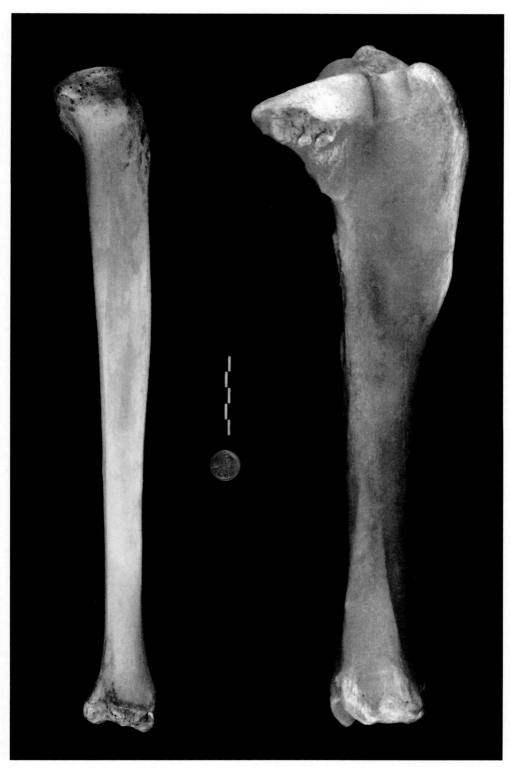

Fig. 2-10. A human right tibia (lateral view) is compared to a horse's right tibia (lateral view).

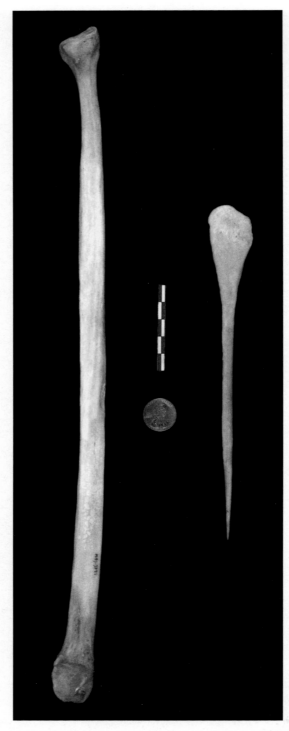

Fig. 2-11. A human right fibula (medial view) is compared to a horse's right fibula (lateral view). The horse's fibula is greatly reduced. The rounded head is transversely flattened, and the shaft tapers to a point.

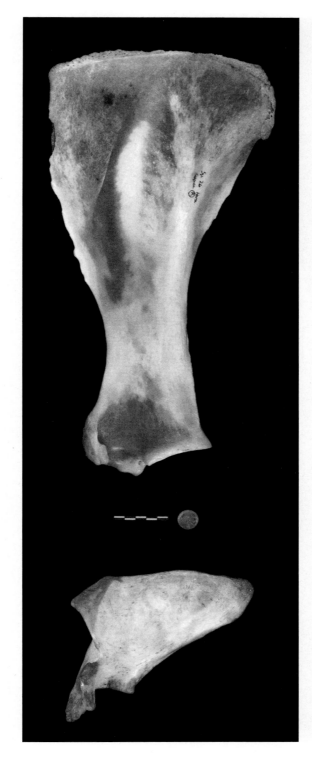

Fig. 2-12. A human right scapula (anterior view) is compared to a horse's right scapula (medial view). Both scapulae are oriented as they would be in a human.

21

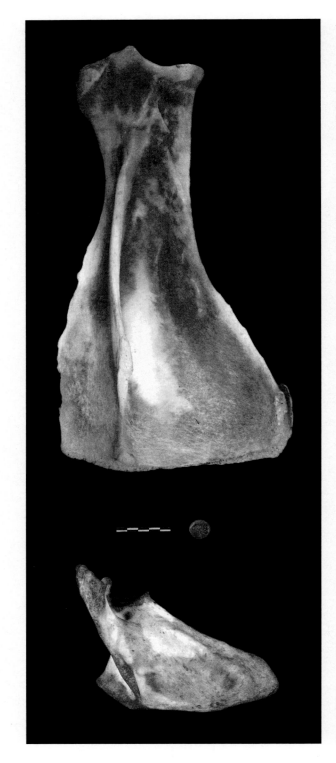

Fig. 2-13. A human right scapula (posterior view) is compared to a horse's right scapula (lateral view). Note that the spine of the horse's scapula rises from the scapular neck.

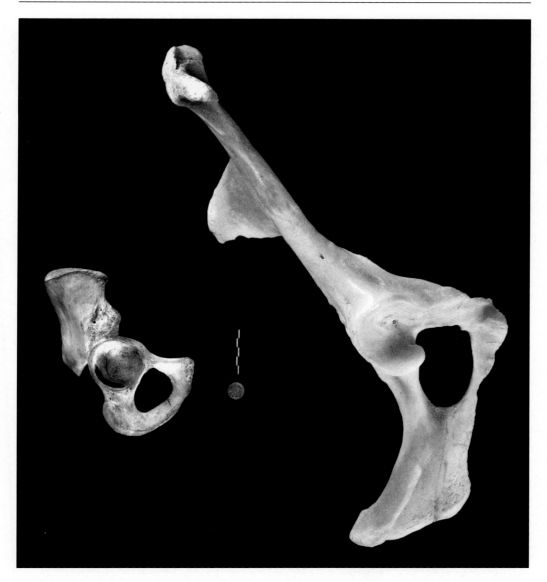

Fig. 2-14. A human right innominate (lateral view) is compared to a horse's right innominate (lateral view). The articular surface on the horse's acetabulum is crescent-shaped.

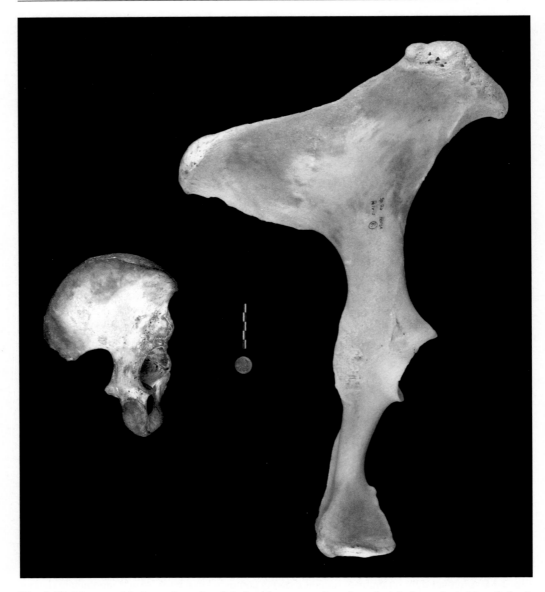

Fig. 2-15. A human right innominate (medial view) is compared to a horse's right innominate (dorsal view).

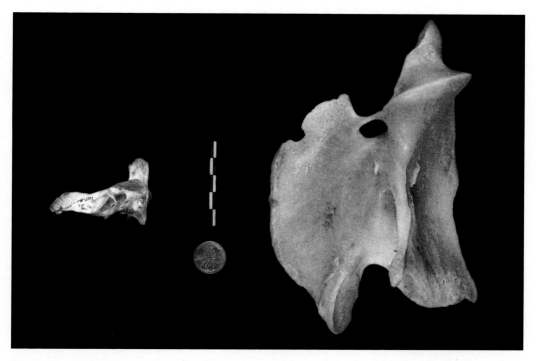

Fig. 2-16. A human axis (C2) is compared to a horse's axis (C2). Both views are lateral. The cervical vertebrae generally reflect the length of the animal's neck. Note how much longer the horse's axis is when compared to the human axis.

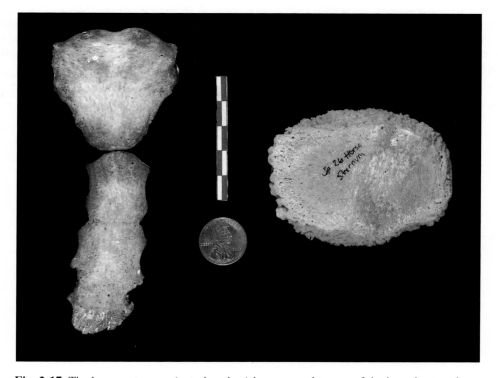

Fig. 2-17. The human sternum (anterior view) is compared to one of the horse's sternabrae.

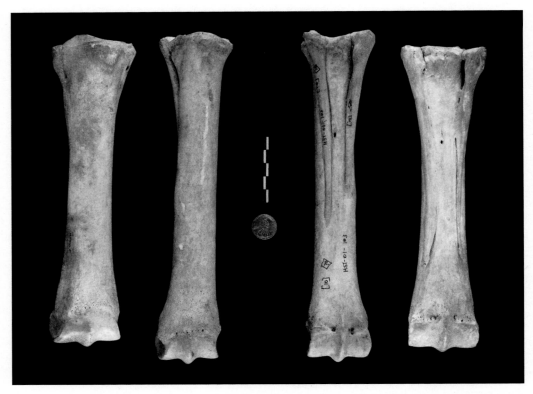

Fig. 2-18. A horse's right metacarpus and metatarsus (dorsal views) are shown on the left. The horse's right metacarpus (volar view) and right metatarsus (plantar view) are shown on the right. The horse has a single main metacarpus (3rd metacarpal) and metatarsus (3rd metatarsal). The remnants of the 2nd and 4th metacarpals and 2nd and 4th metatarsals can be seen volar/plantar views (shown on the right). These "splint bones" (lateral metapodia) taper to a point about half way down the shaft of the main metapodial.

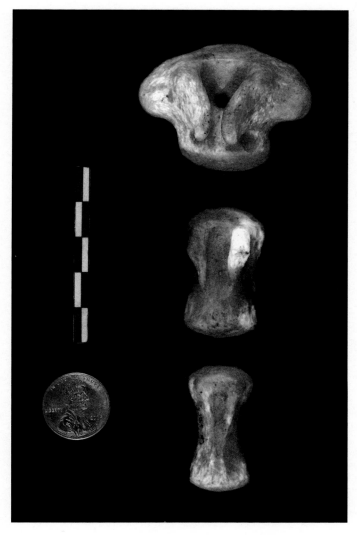

Fig. 2-19. Three caudal, or tail, vertebrae of a horse (dorsal views). While the numbers may vary, most horses have about 18 caudal vertebrae. These bones can sometime be confused with human phalanges.

3   Human vs Cow

Fig. 3-00. Cow skull (basal view). The cow has no upper incisors or canines. The tooth cow's maxillary tooth row includes three premolars and three molars on each side of the maxilla. The mandibular dental formula is 3.1.3.3.

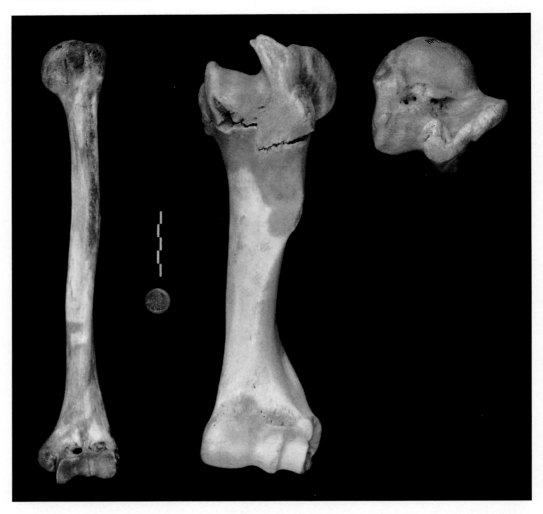

Fig. 3-01. Human left humerus (anterior view) compared to cow's left humerus (cranial view). The proximal epiphysis of the cow's humerus is shown on the right. On the proximal end, the greater and less tubercles are far more well-developed in the cow than they are in the human. On this distal end, the cow has a barrel-shaped trochlea.

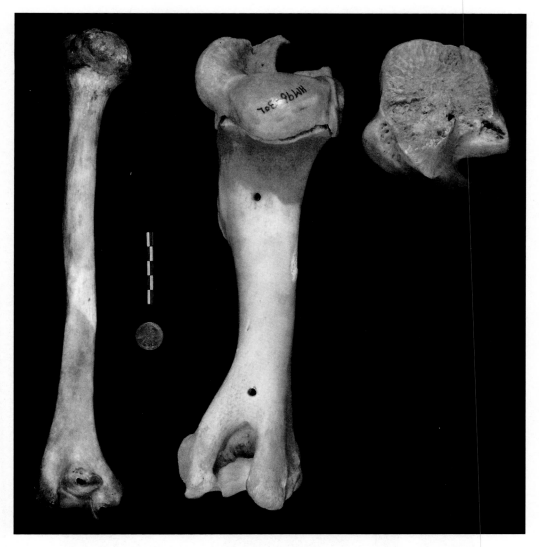

Fig. 3-02. Human left humerus (posterior view) compared to cow's left humerus (caudal view). The cow's unfused proximal epiphysis is shown on the right. Note that the cow humerus has a particularly deep olecranon fossa when compared to the human example.

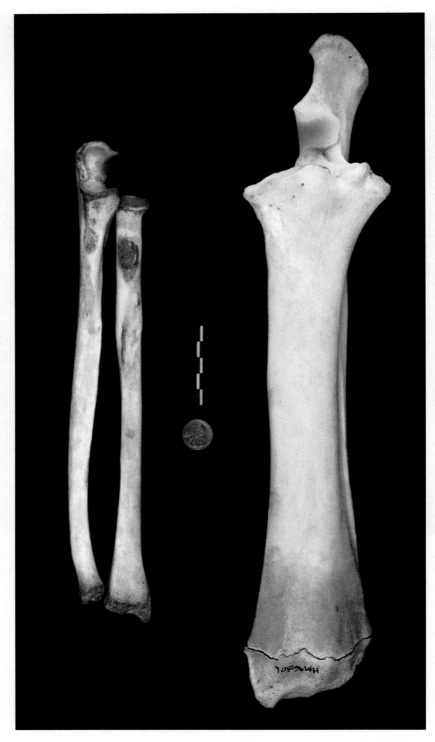

Fig. 3-03. Human left radius and ulna (anterior views) compared to a cow's left radius and ulna (cranial view). The human radius and ulna are roughly equal in size. With the exception of the large olecranon process, the cow's ulna is greatly reduced. While the human radius has a distinct head, the cow's radius has a broad and slightly concave articular surface.

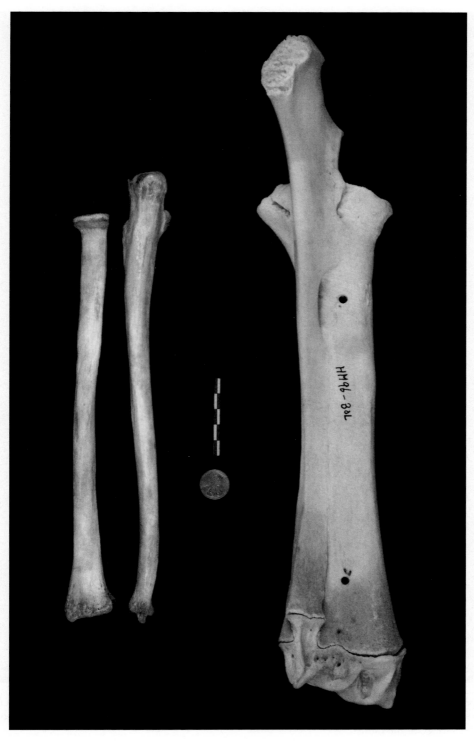

Fig. 3-04. Human left radius and ulna (posterior views) compared to the cow's left radius and ulna (caudal view). Note that the cow's radius is fused to the lateral side of the radius shaft.

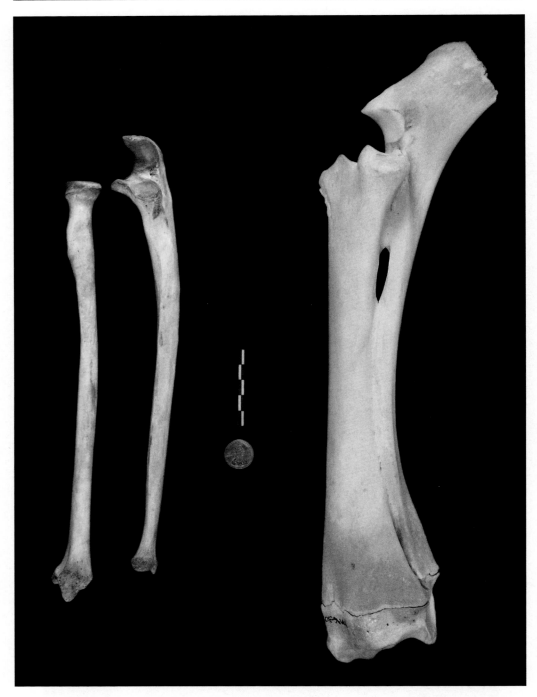

Fig. 3-05. Human left radius and ulna (lateral views) compared to the cow's left radius and ulna (lateral views). The shaft of the cow's radius is quite slender. In adult animals, it is usually fused to the caudal surface of the radius.

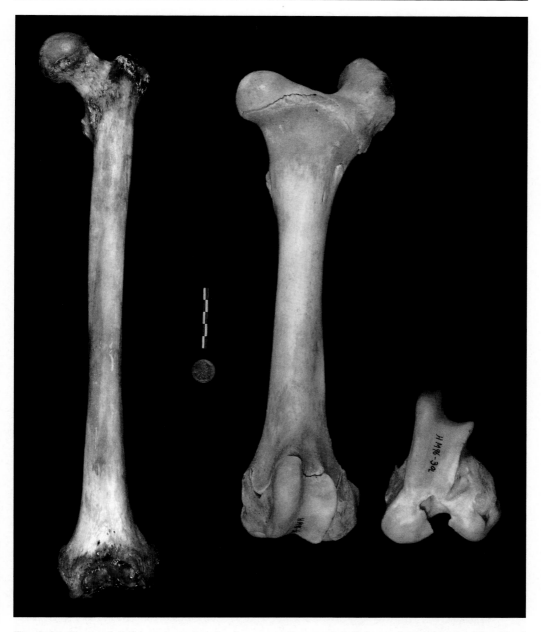

Fig. 3-06. Human left femur (anterior view) compared to cow's left femur (cranial view). The distal epiphysis of the cow's femur is shown separately to the right. The cow skeleton illustrated here is a 6–7 year old ox, or male castrate. While the distal epiphysis usually fuses by about 3.5–4 years of age (Silver 1969) castration has delayed epiphyseal fusion in this specimen. The greatest length of the human femur extends from the head to the distal condyles; in the cow the greatest length extends from the greater trochanter to the condyles.

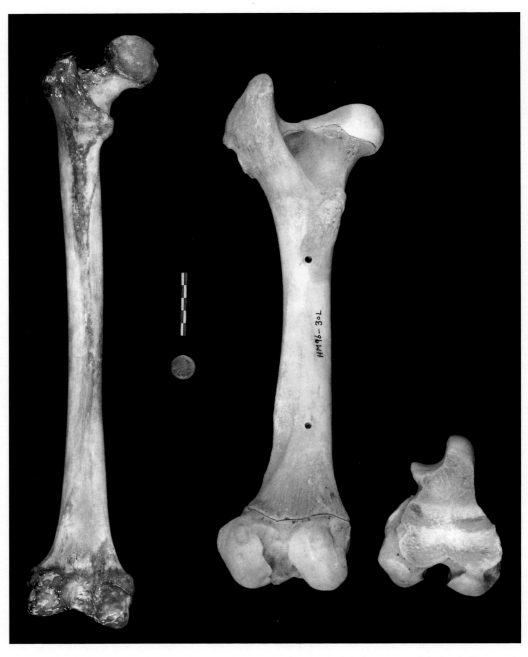

Fig. 3-07. Human left femur (posterior view) compared to cow's left femur (caudal view). The cow's unfused distal epiphysis is shown on the right. The human femur includes a linea aspera which appears as a projecting ridge for the muscle attachment. This is a unique human feature that is not seen in other mammals as it is an area of attachment for the muscles used in bipedalism. The cow's femur includes a supercondular fossa that can be seen on the lateral portion of the shaft.

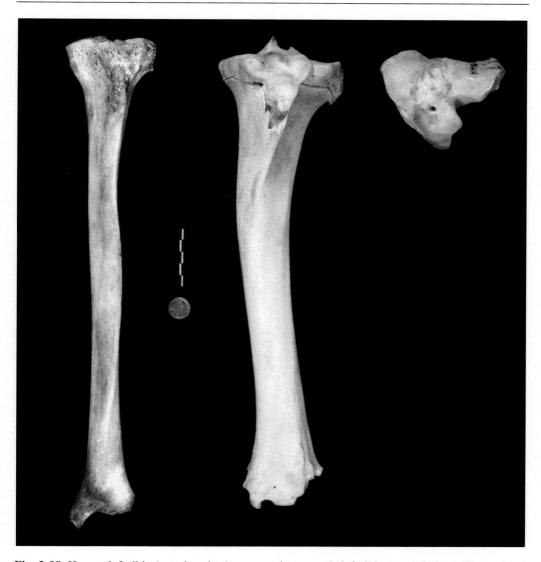

Fig. 3-08. Human left tibia (anterior view) compared to a cow's left tibia (cranial view). The unfused proximal epiphysis of the cow's tibia is shown at right. The shaft of the cow's tibia is significantly more robust. The distal end of the cow's tibia includes two parallel articular facets for articulation with the astragalus.

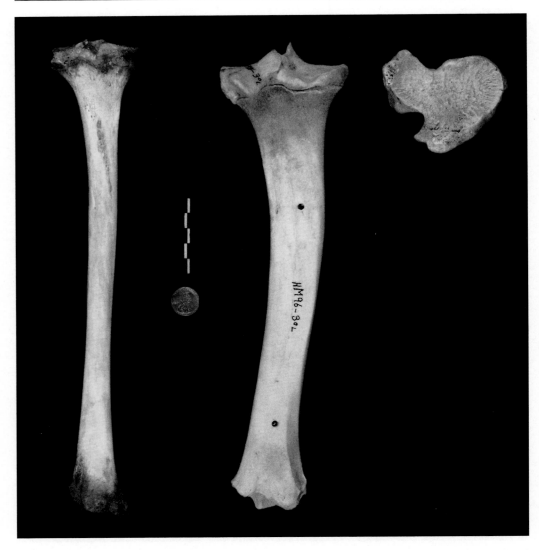

Fig. 3-09. Human left tibia (posterior view) compared to the cow's tibia (caudal view). The unfused proximal epiphysis of the cow's tibia is shown at right.

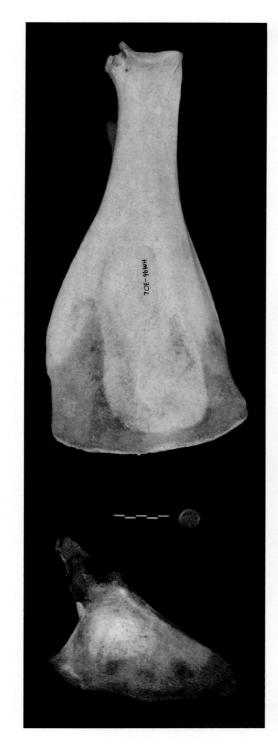

Fig. 3-10. Human left scapula (anterior view) compared to the cow's left scapula (medial view). Note that these bones are oriented as they would be in the human skeleton. Since the cow is a quadruped, the glenoid cavity forms the distal part of the bone. The blade of the cow's scapula is shaped like an elongated triangle.

39

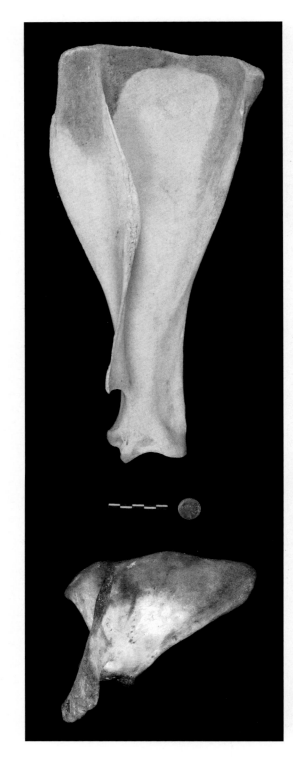

Fig. 3-11. Human left scapula (posterior view) compared to the cow's left scapula (lateral view). The acromion process is well developed in humans; it is very small in the cow and other ruminant artiodactyls. The greatest length of the cow's scapula (and the scapulas of most other mammals) extends from the glenoid cavity to the dorsal border (parallel to the scapular spine). This is not the case in humans. In humans, the greatest length lies between the superior and inferior borders.

40

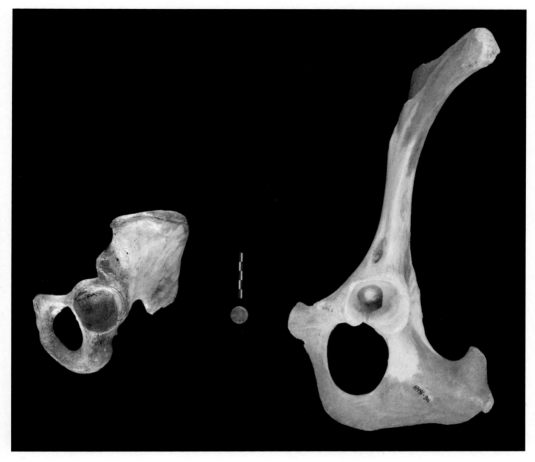

Fig. 3-12. A human left innominate (lateral view) is compared to cow's left innominate (lateral view).

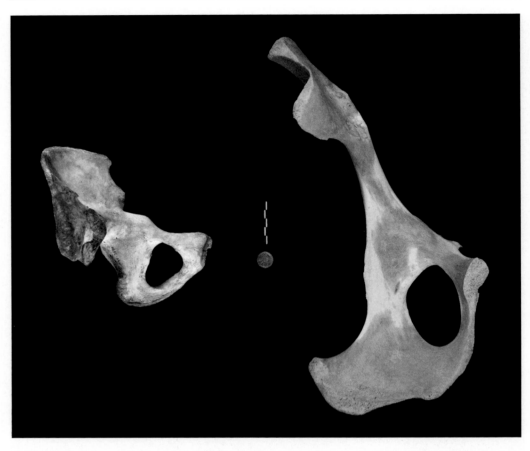

Fig. 3-13. A human left pelvis (medial view) is compared to a cow's left pelvis (medial view).

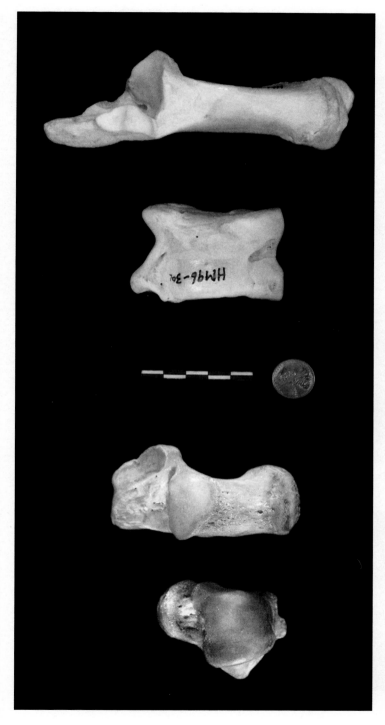

Fig. 3-14. Human left talus and calcaneus (superior views) compared to cow's left astragalus (plantar view) and calcaneus (dorsal view). The human talus has a distinctive head, while the cow's astragalus has the "double pulley" form that is typical of all artiodactyls (even-toed ungulates). The cow's calcaneus is elongated, and the dorsal surface includes an articular facet for the malleolus, a small tarsal bone that is the evolutionary remnant of the distal fibula. The cow does not have a separate fibula.

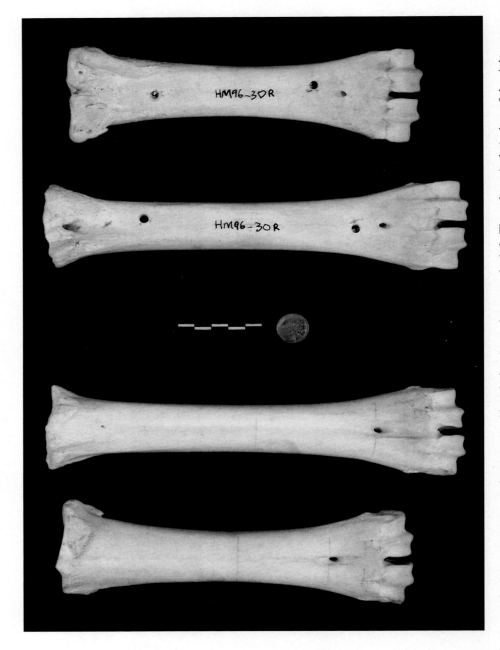

Fig. 3-15. The cow's right metacarpus and metatarsus (dorsal views) are shown on the left. The palmar (volar) view of the right metacarpus and the plantar view of the right metatarsus are shown on the right. These bones are the formed through the fusion of the third and fourth metacarpals and metatarsals. Each of the two distal condyles articulates with a first phalanx.

44

4   Human vs Bear

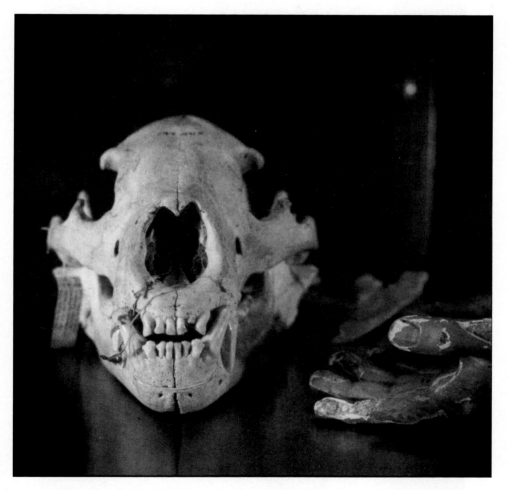

Fig. 4-00. A bear's cranium and mandible (cranial view). The bear's upper dentition includes three incisors, one canine, between two and four premolars, and two molars. The mandibular dentition includes three incisors, one canine, two to four premolars, and three molars.

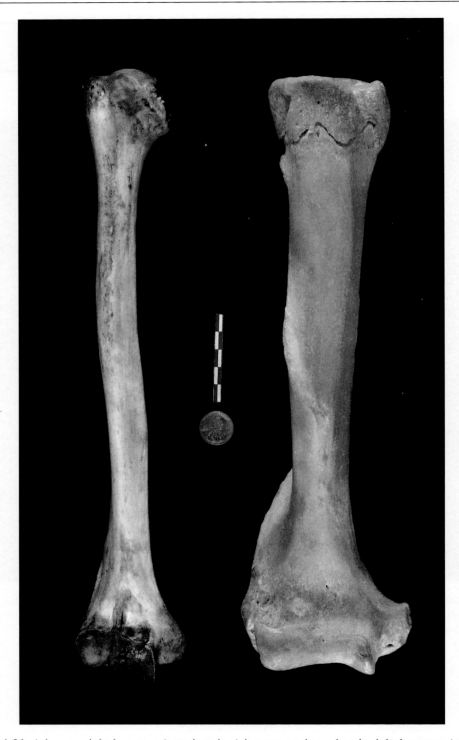

Fig. 4-01. A human right humerus (anterior view) is compared to a bear's right humerus (cranial view). The lateral epicondylar crest (proximal to the lateral epicondyle) is well developed in the bear, as is the deltoid tuberosity.

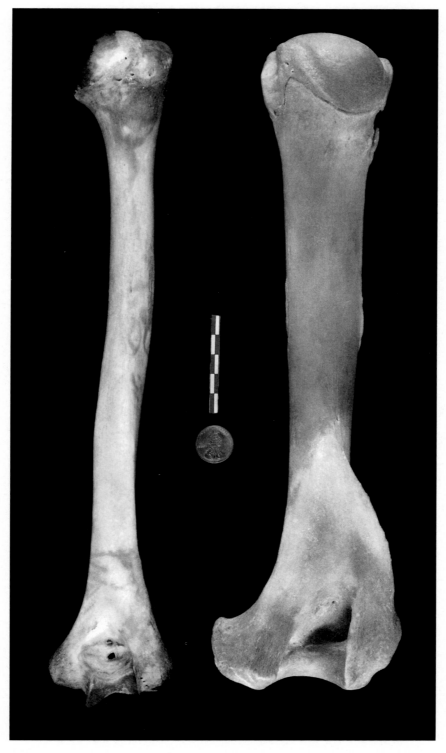

Fig. 4-02. A human right humerus (posterior view) is compared to a bear's right humerus (caudal view).

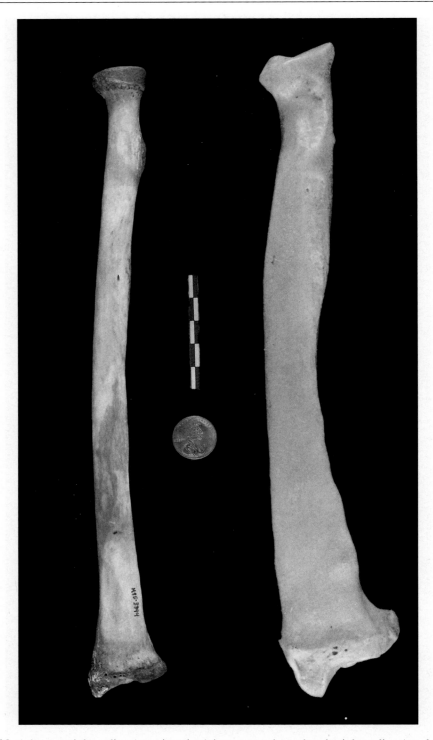

Fig. 4-03. A human right radius (anterior view) is compared to a bear's right radius (caudal view).
Human skeletons are oriented with the palms forward, so that the radius and ulna are not crossed.
Quadrupedal animals are oriented with their paws facing the ground. This means that the proximal
ulna is medial to the radius, while the distal ulna is on the lateral side.

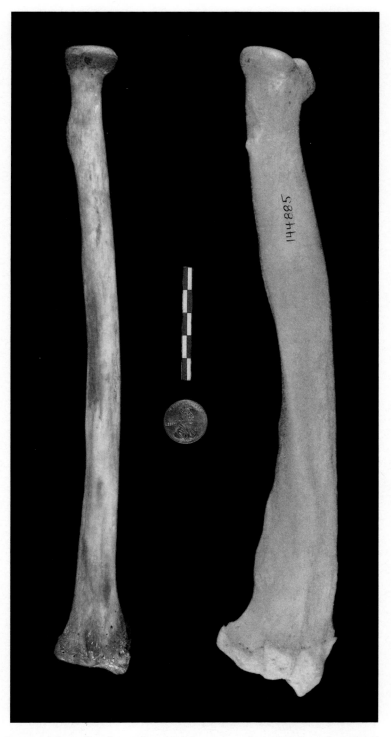

Fig. 4-04. A human right radius (posterior view) is compared to a bear's right radius (cranial view).

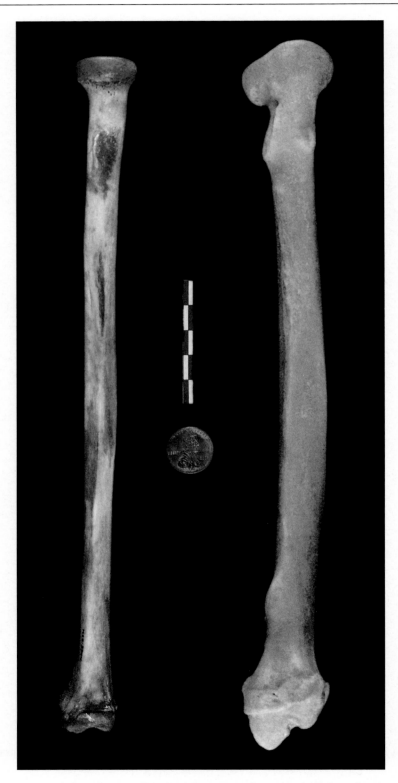

Fig. 4-05. A human right radius (medial view) is compared to a bear's right radius (medial view).

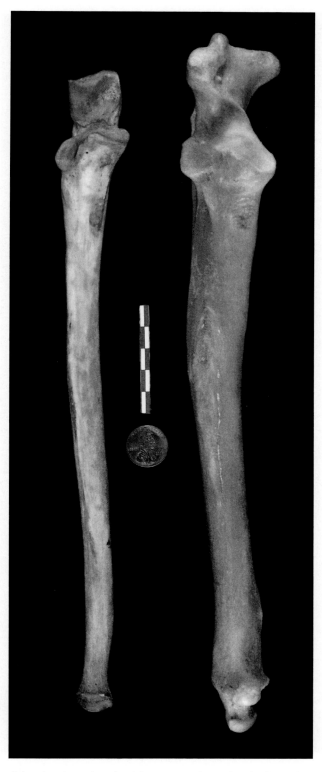

Fig. 4-06. A human right ulna (anterior view) is compared to a bear's right ulna (cranial view). Note that the bear's olecranon process is larger and more well-developed.

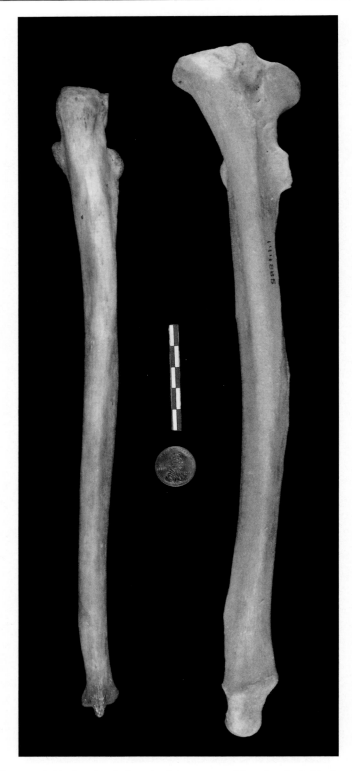

Fig. 4-07. A human right ulna (posterior view) is compared to a bear's right ulna (caudal view).

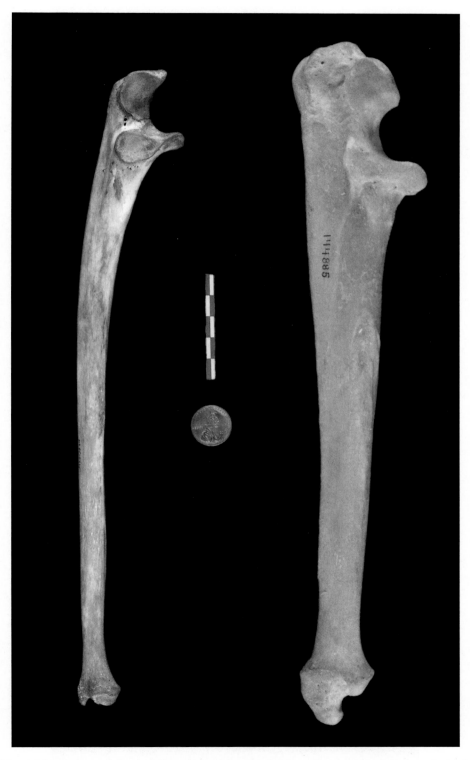

Fig. 4-08. A human right ulna (lateral view) is compared to a bear's right ulna (lateral view).

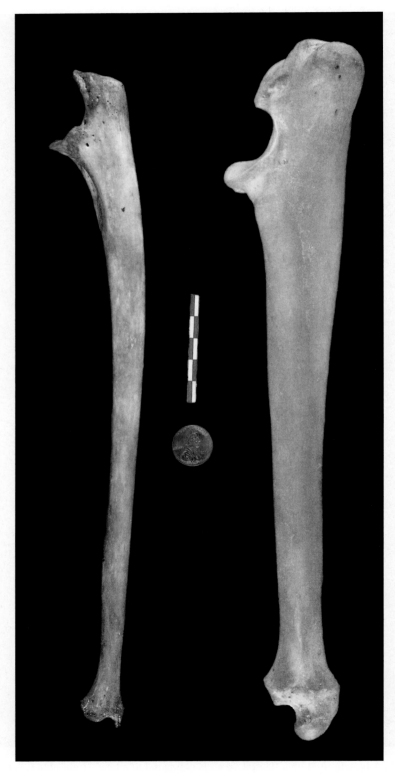

Fig. 4-09. A human right ulna (medial view) is compared to a bear's right ulna (medial view).

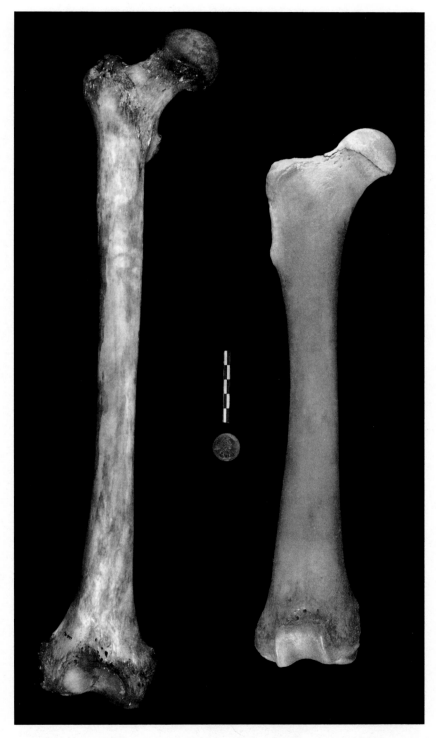

Fig. 4-10. A human right femur (anterior view) is compared to a bear's right femur (cranial view). The distal condyles of the human femur show a distinctive asymmetry that is a result of "kneeing in" or bringing the knees under the body. This is also referred to as the valgus knee.

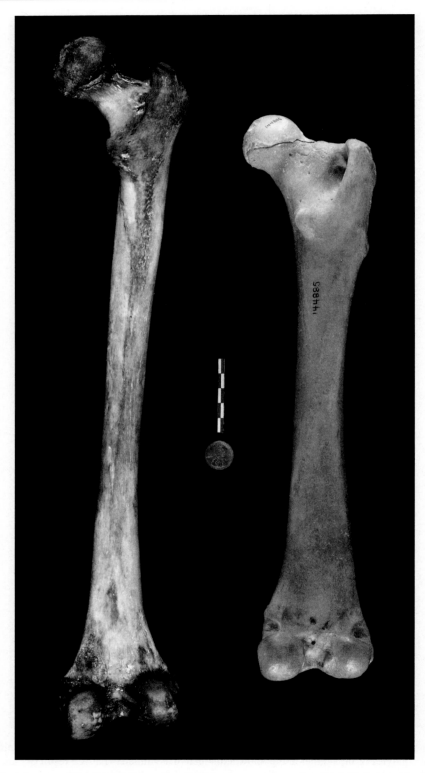

Fig. 4-11. A human right femur (posterior view) is compared to a bear's right femur (caudal view). The human femur shows the distinctive linea aspera which is not seen in quadrupeds.

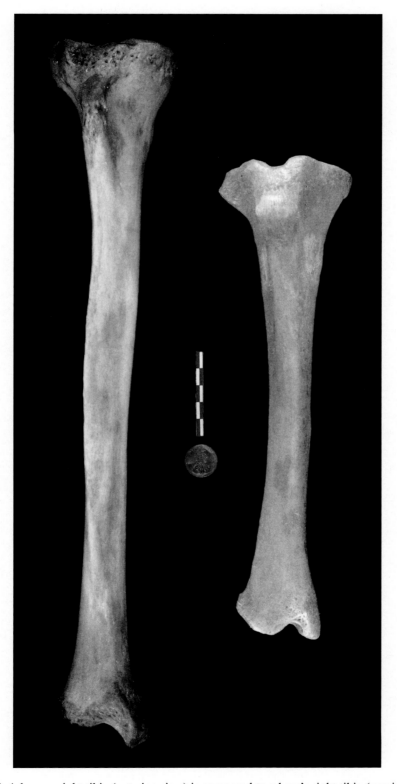

Fig. 4-12. A human right tibia (anterior view) is compared to a bear's right tibia (cranial view).

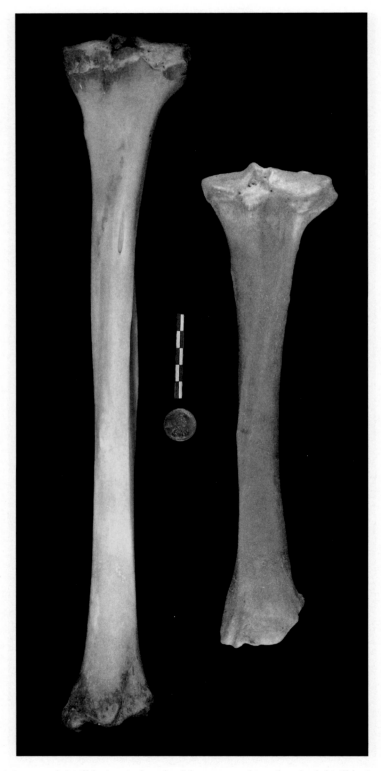

Fig. 4-13. A human right tibia (posterior view) is compared to a bear's right tibia (caudal view).

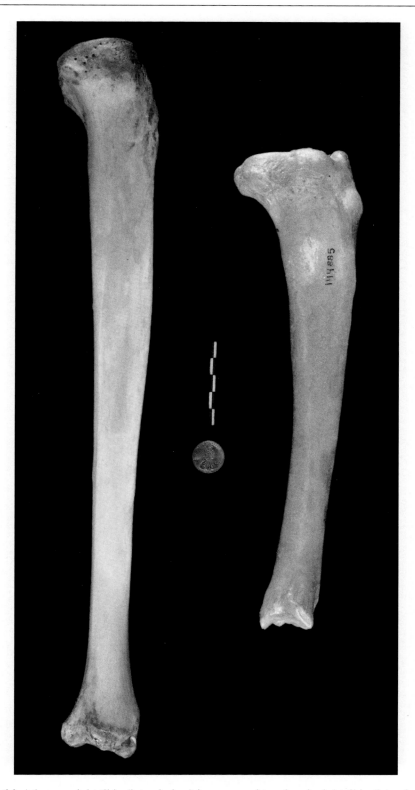

Fig. 4-14. A human right tibia (lateral view) is compared to a bear's right tibia (lateral view).

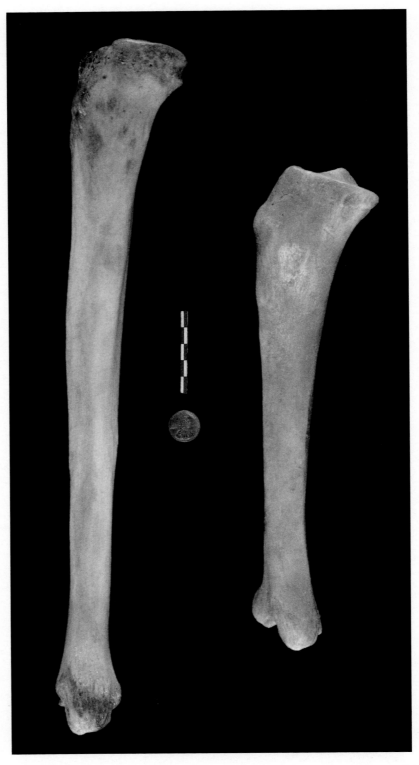

Fig. 4-15. A human right tibia (medial view) is compared to a bear's right tibia (medial view).

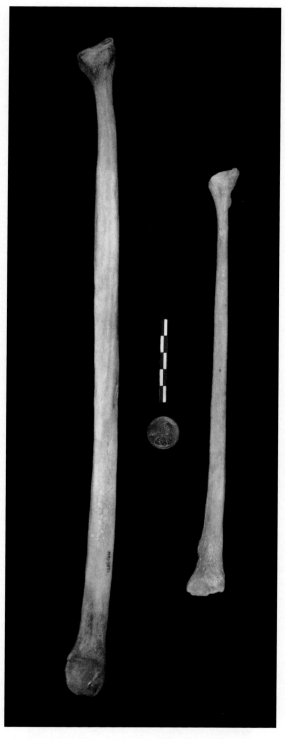

Fig. 4-16. A human right fibula (medial view) is compared to a bear's right fibula (medial view).

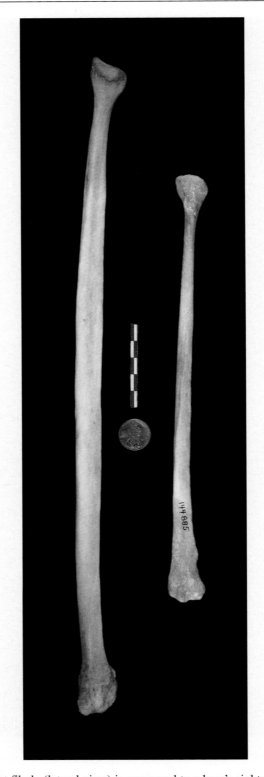

Fig. 4-17. A human right fibula (lateral view) is compared to a bear's right fibula (lateral view).

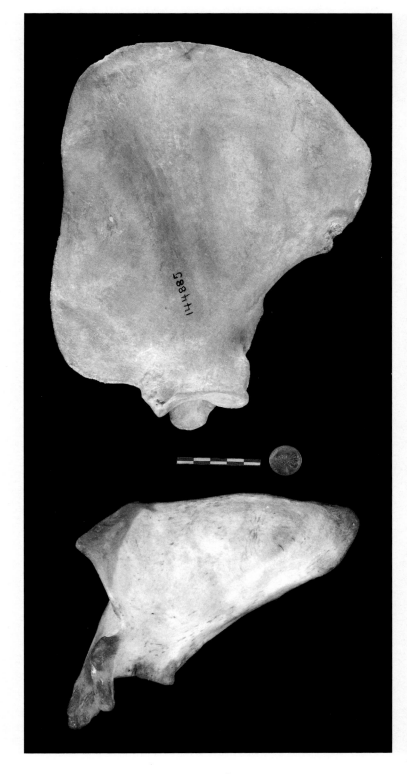

Fig. 4-18. A human scapula (anterior view) is compared to a bear's scapula (medial view). Both scapulae are oriented as they would be in a human skeleton.

63

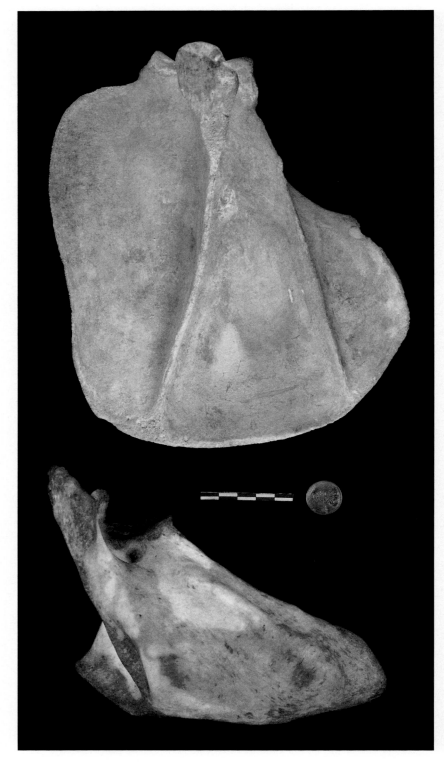

Fig. 4-19. A human scapula (posterior view) is compared to a bear's scapula (lateral view). The spine of the bear's scapula divides the scapula into roughly equal halves.

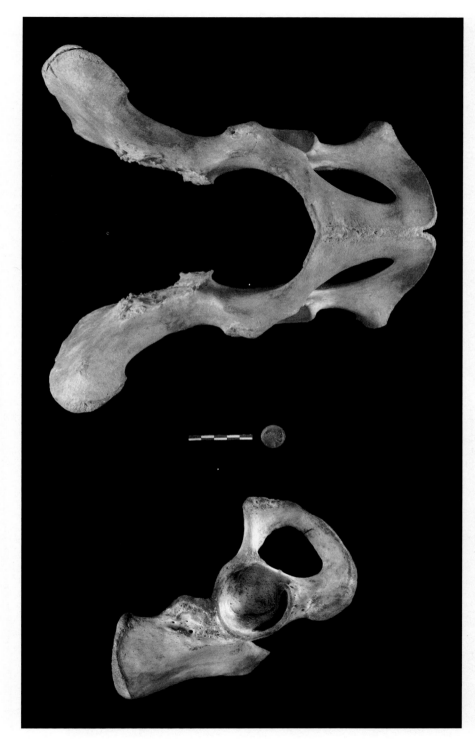

Fig. 4-20. A human right innominate (lateral view) is compared to a bear's pelvis (ventral view). In the bear, the two innominates have fused along the pubic symphysis. The broad ilium is typical of human pelves and is related to bipedalism. Quadrupeds have longer, narrower ilia.

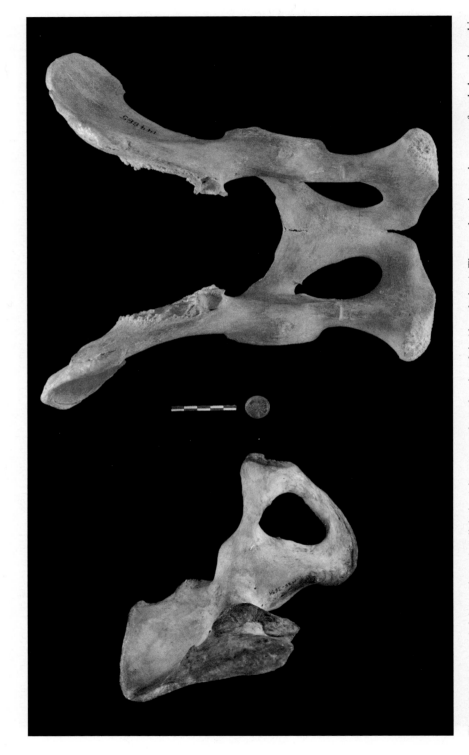

Fig. 4-21. A human left innominate (medial view) is compared to a bear's pelvis (dorsal view). The two bear innominates are fused along the pubic symphysis.

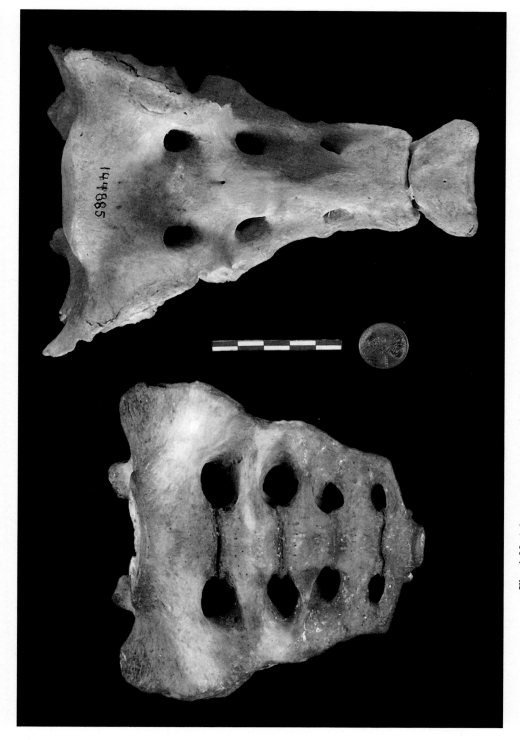

Fig. 4-22. A human sacrum (anterior view) is compared to a bear's sacrum (ventral view).

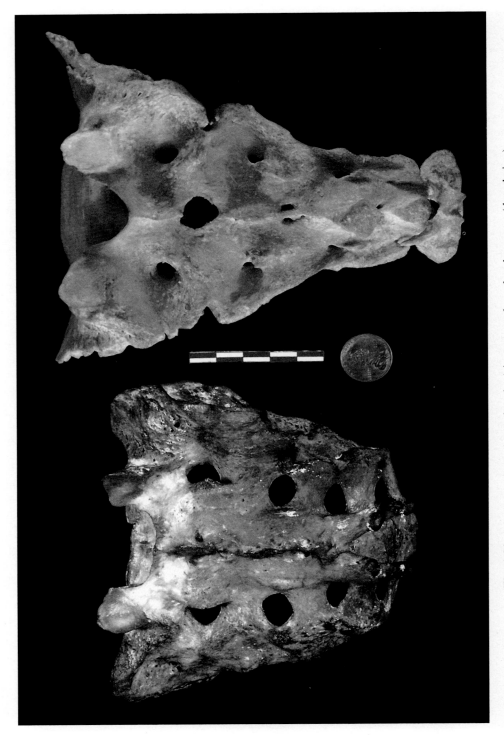

Fig. 4-23. A human sacrum (posterior view) is compared to a bear's sacrum (dorsal view).

68

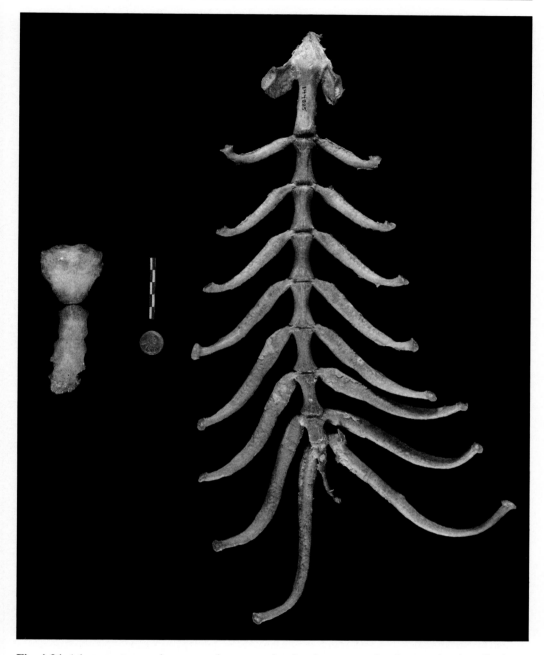

Fig. 4-24. A human sternum is compared to a complete bear's sternum plus the associated cartilaginous ribs. Note the differences in the shape of the human and the bear's manubrium.

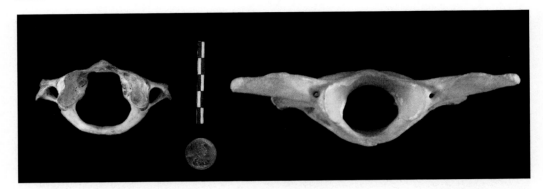

Fig. 4-25. Human atlas (C1, superior view) is compared to a bear atlas (C1, cranial view). Note that large wings on the bear atlas; these are typical of carnivores.

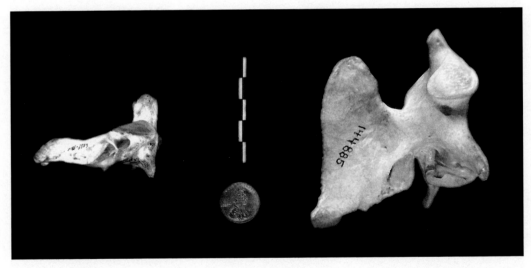

Fig. 4-26. Human axis (C2, lateral view) is compared to a bear atlas (C2, lateral view). The bear atlas is much longer and includes a large spinous process.

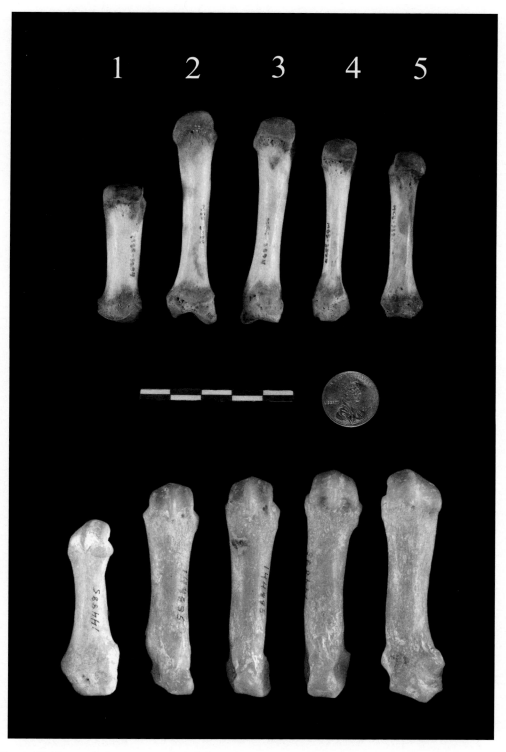

Fig. 4-27. Human left metacarpals 1-5 (anterior view) are compared to a bear's left metacarpals (volar view). The similarity in size and shape between human and bear metacarpals is frequently noted.

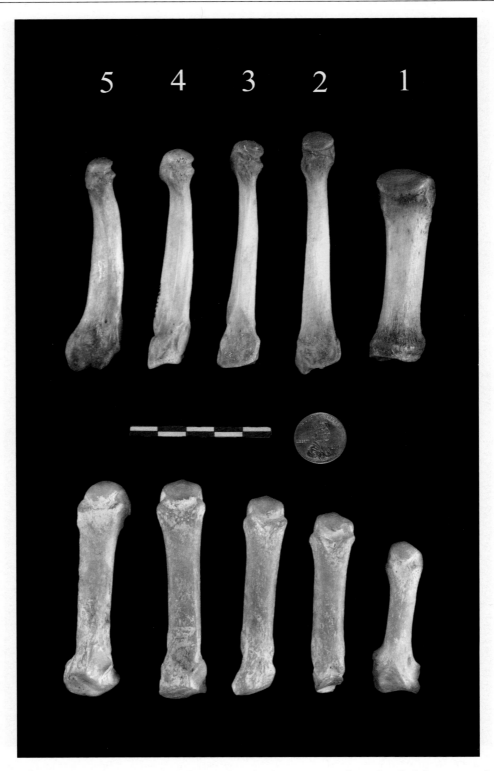

Fig. 4-28. Human left metatarsals 1-5 (superior view) are compared to a bear's left metatarsals (dorsal view).

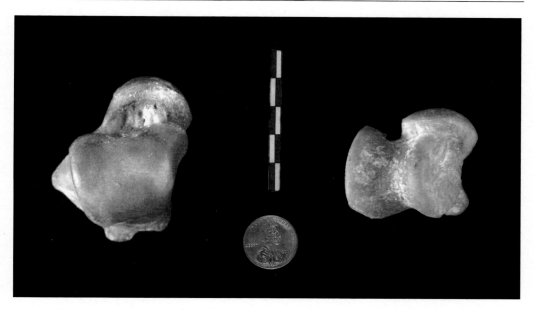

Fig. 4-29. A human left talus (superior view) is compared to a bear's left astragalus (plantar view).

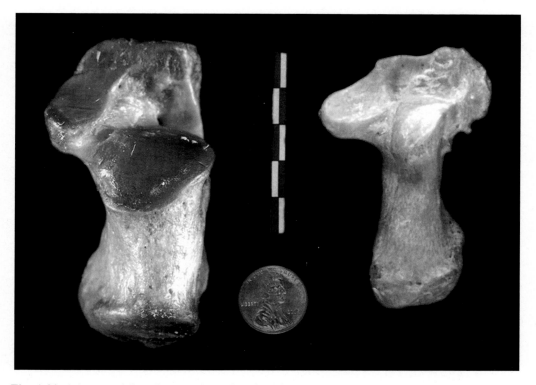

Fig. 4-30. A human right calcaneus (superior view) is compared to a bear's right calcaneus (dorsal view). While these two bones are generally quite similar in form, note the differences in the shape of the sustentaculum tali.

5   Human vs Deer

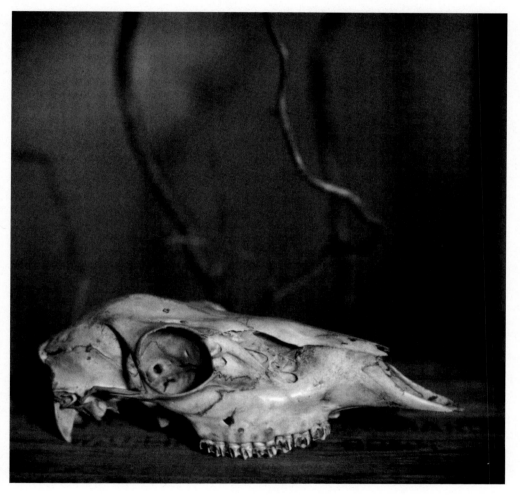

Fig. 5-00a. A cranium of a female white-tailed deer. In most deer species, the males have antlers while the females do not. The one exception is the reindeer; both male and female reindeer have antlers.

Fig. 5-00b. A cranium of a male white-tailed deer. The dental formula for the upper dentition includes no incisors and canines, three premolars, and three molars. On the mandible, the dental formula is 3.1.3.3.

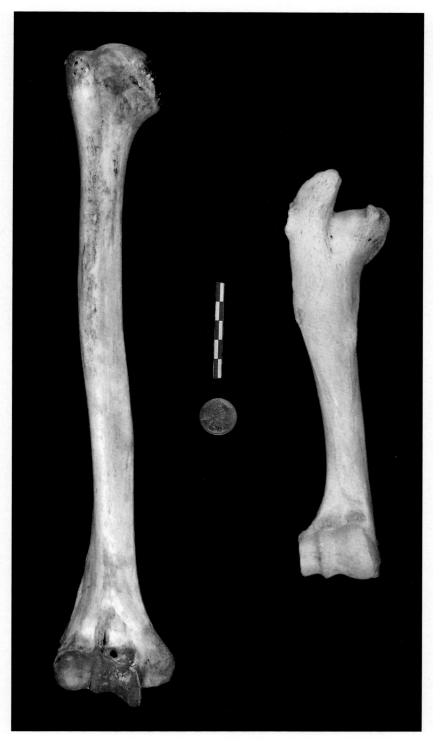

Fig. 5-01. A human right humerus (anterior view) is compared to a deer's right humerus (cranial view). Note the height of the greater tubercle on the deer's proximal humerus.

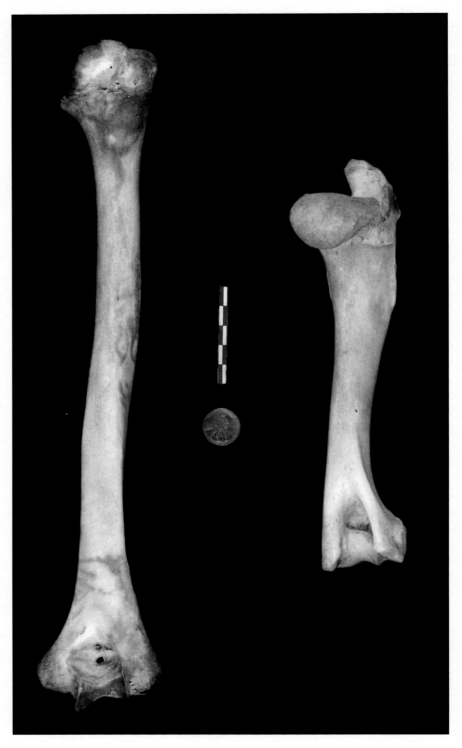

Fig. 5-02. A human right humerus (posterior view) is compared to a deer's right humerus (caudal view). Note the deeper olecranon fossa on the deer's distal humerus.

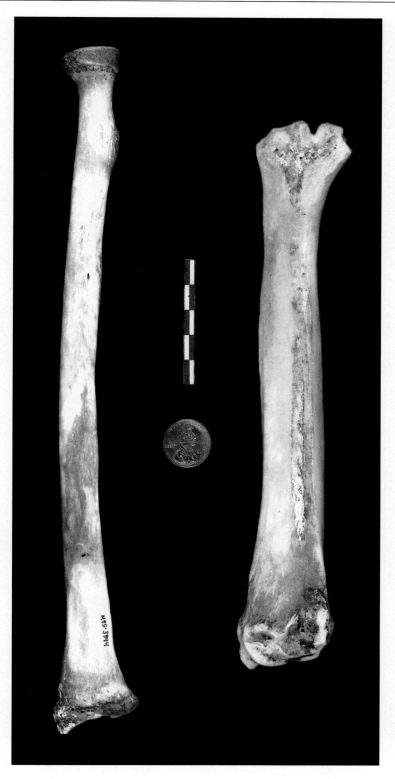

Fig. 5-03. A human right radius (anterior view) is compared to a deer's right radius (caudal view). The articulation with the ulna is visible on the lateral part of the caudal surface of the deer's radius.

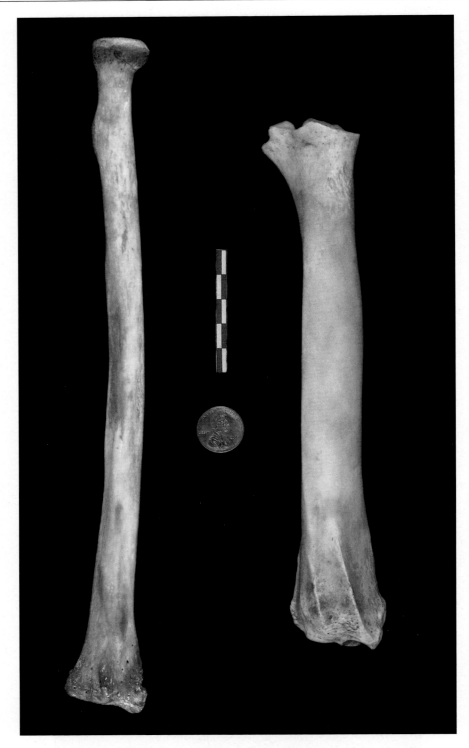

Fig. 5-04. A human right radius (posterior view) is compared to a deer's right radius (cranial view). The human proximal radius has a distinct head and neck, while the deer radius has a slightly concave articular surface.

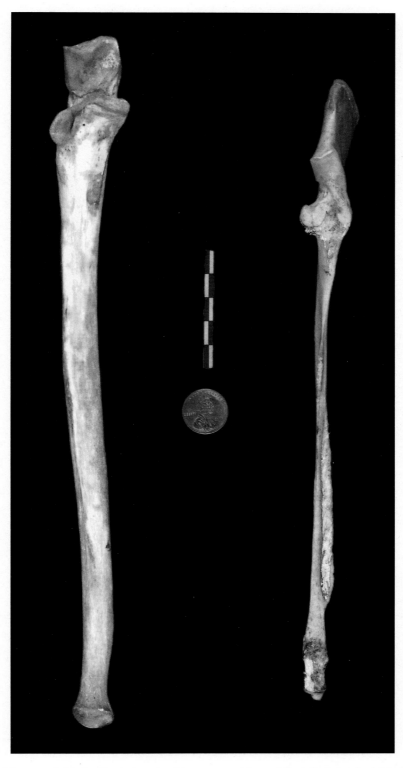

Fig. 5-05. A human right ulna (anterior view) is compared to a white-tailed deer's right ulna (cranial view). Note the slender shaft of the deer's ulna.

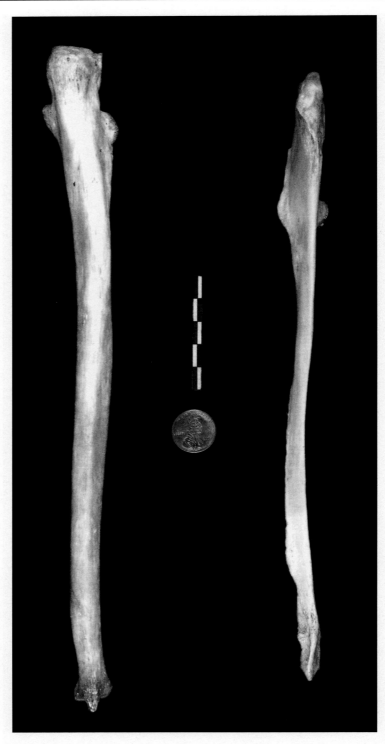

Fig. 5-06. A human right ulna (posterior view) is compared to a white-tailed deer's right ulna (caudal view).

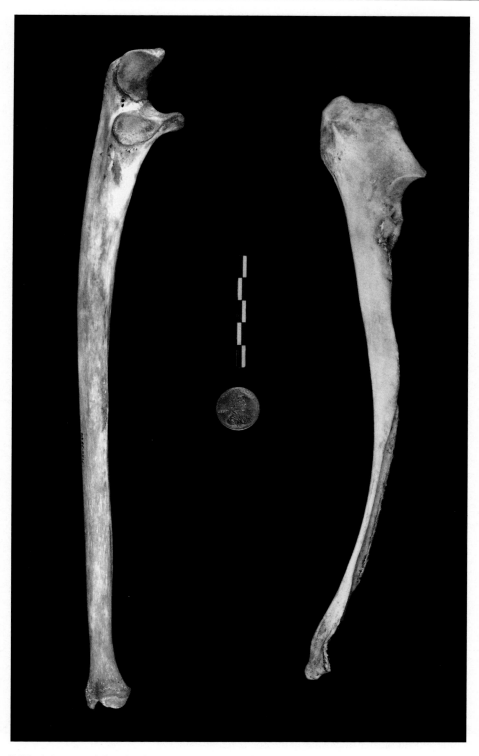

Fig. 5-07. A human right ulna is compared to a white-tailed deer's right ulna (both lateral views). Note the well-developed olecranon process on the deer's ulna.

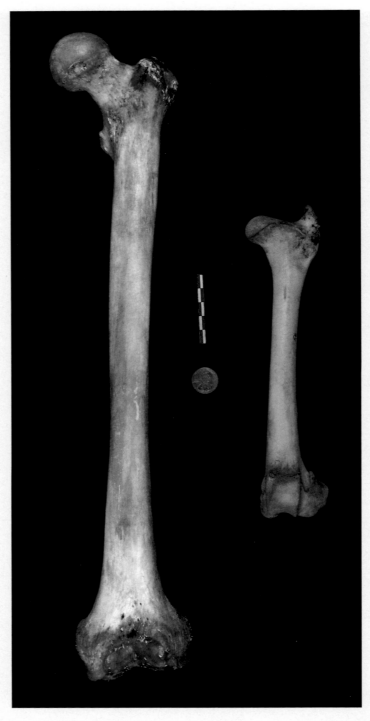

Fig. 5-08. A human left femur (anterior view) is compared to a deer's left femur (cranial view). Note the well-developed greater trochanter in the deer's femur.

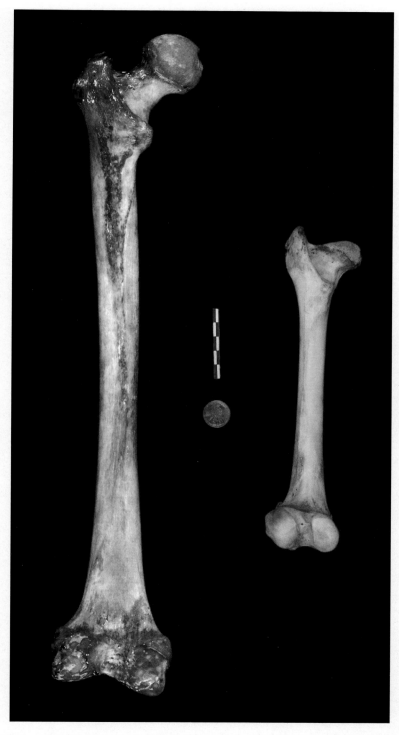

Fig. 5-09. A human left femur (posterior view) is compared to a deer's left femur (caudal view). The linea aspera is visible on the human left femur. This is a distinctly human feature that is associated with the muscles used in bipedalism.

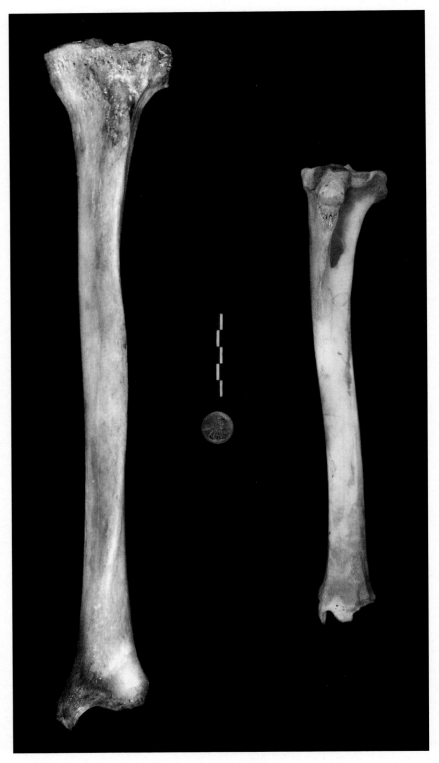

Fig. 5-10. A human left tibia (anterior view) is compared to a deer's left tibia (cranial view). The deer's distal tibia includes two parallel articular surfaces for articulation with the astragalus.

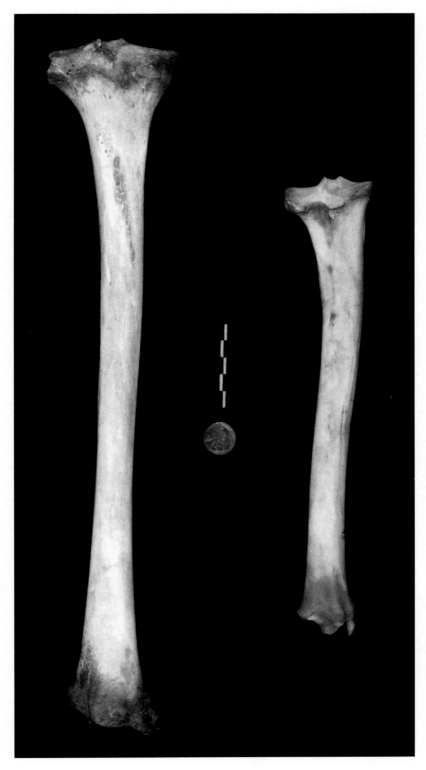

Fig. 5-11. A human left tibia (posterior view) is compared to a deer's left tibia (caudal view).

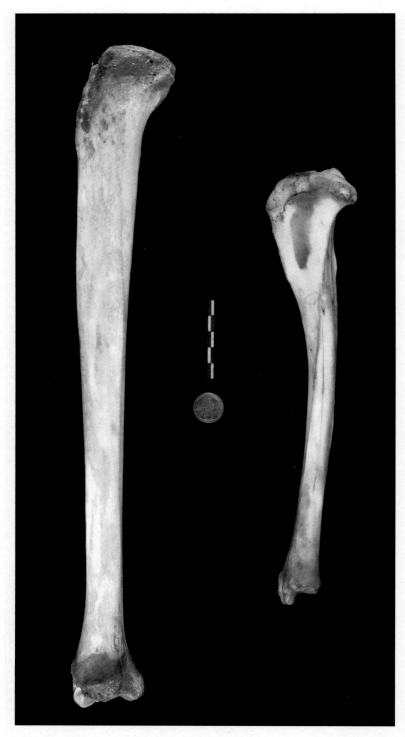

Fig. 5-12. A human left tibia (lateral view) is compared to a deer's left tibia (lateral view). Note the well-developed tibial tuberosity on the deer's tibia.

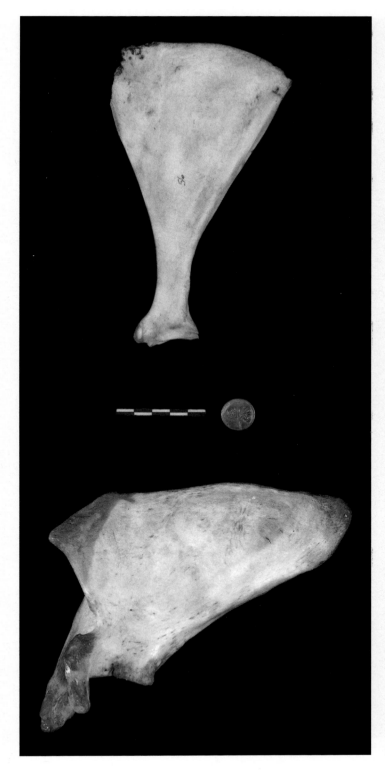

Fig. 5-13. A human right scapula (anterior view) is compared to a deer's right scapula (medial view). Both bones are positioned as they would be in a human.

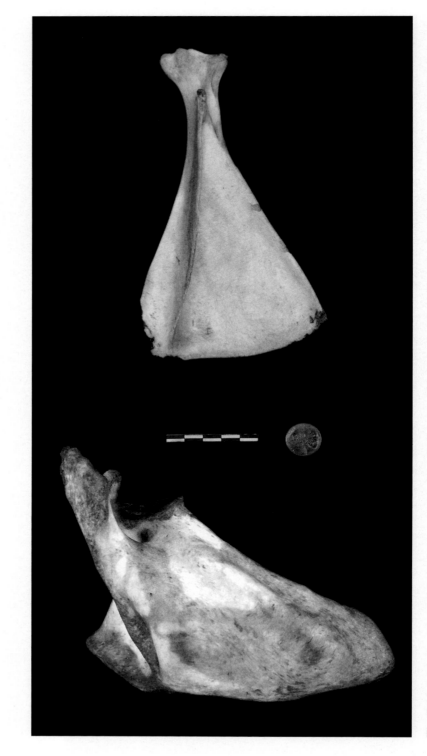

Fig. 5-14. A human right scapula (posterior view) is compared to a deer's right scapula (lateral view). Note the large acromion process on the human scapula. The glenoid cavity on the deer's scapula is very round in shape (not pictured).

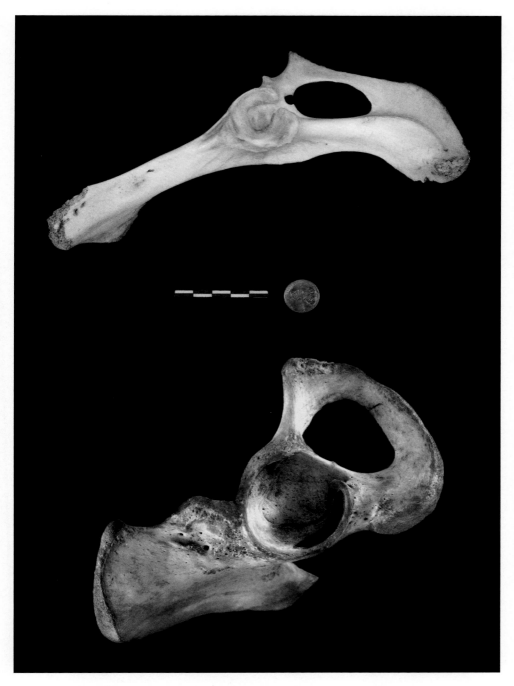

Fig. 5-15. A human right innominate (lateral view) is compared to a deer right innominate (lateral view).

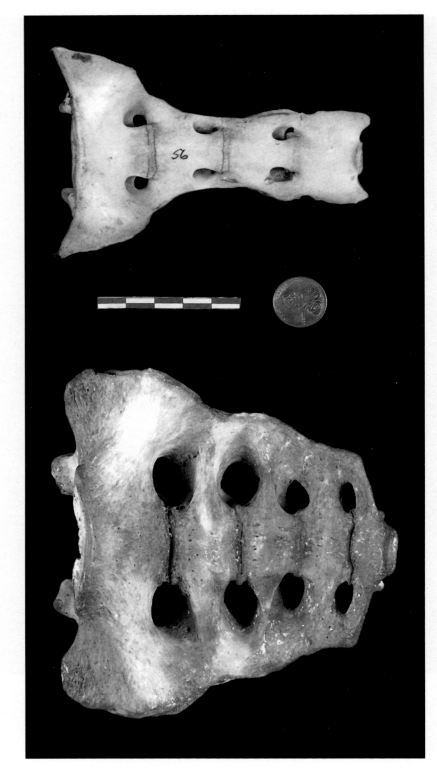

Fig. 5-16. A human sacrum (anterior view) is compared to a deer's sacrum (ventral view). The wings of the deer sacrum are wide in relation to the body.

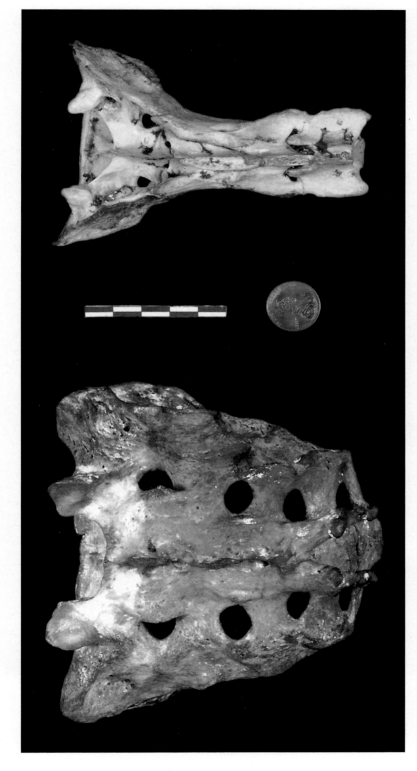

Fig. 5-17. A human sacrum (posterior view) is compared to a deer's sacrum (dorsal view).

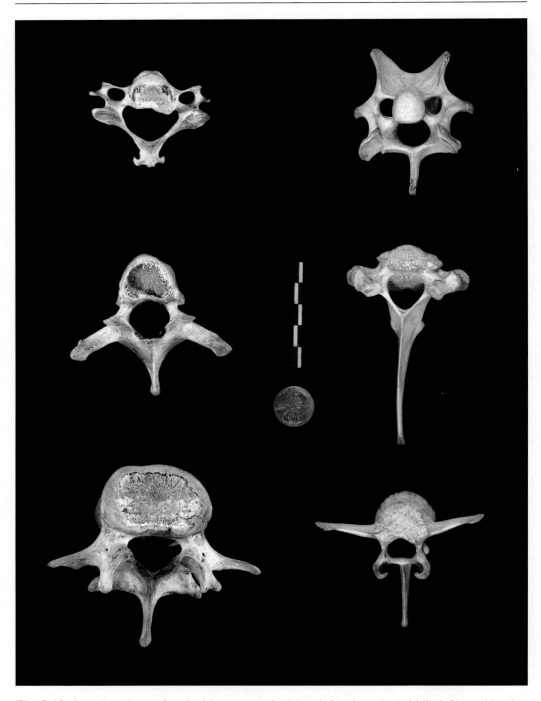

Fig. 5-18. Superior views of typical human cervical (top left), thoracic (middle left), and lumbar (bottom left) vertebrae. Cranial views of typical white-tail deer cervical (top right), thoracic (middle right), and lumbar (bottom right) vertebrae.

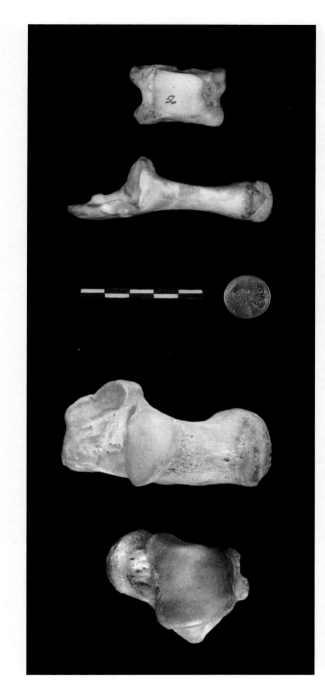

Fig. 5-19. Human left talus and calcaneus (superior views) are compared to a white-tailed deer's left calcaneus (dorsal *view*) and astragalus (plantar *view*). The deer's astragalus has the typical "double-pulley" form that is characteristic of the artiodactyls. The dorsal calcaneus includes an articulation for the malleolus, a small tarsal that is the evolutionary remnant of the distal fibula.

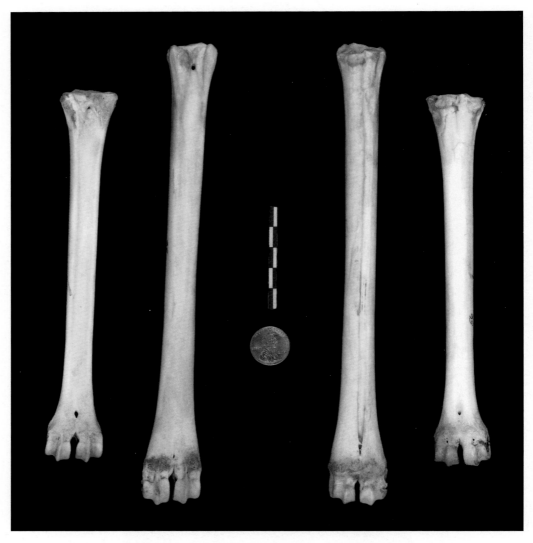

Fig. 5-20. A white-tailed deer's metacarpus (volar view) and metatarsus (plantar view) are shown on the left. The dorsal views of the same elements are shown on the right. These metapodia are composed of the fused third and fourth metacarpals and metatarsals. The deer metapodia have distinctive concave grooves along the plantar/volar surfaces that distinguish them from the metapodia of the bovids.

6   Human vs Pig

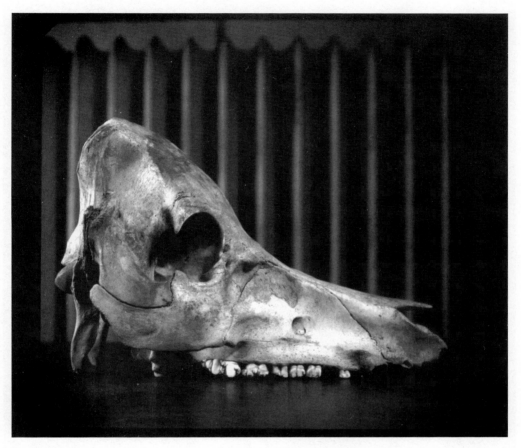

Fig. 6-00. A lateral view of an adult pig cranium. The pig dental formula is 3/3.1/1.4/4.3/3.

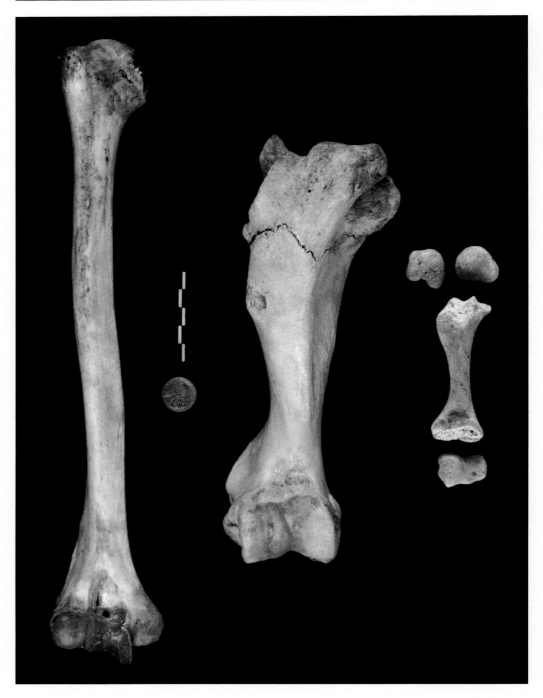

Fig. 6-01. A human right humerus (anterior view) is compared to adult and juvenile pig right humeri (cranial views). Note the large greater tubercle in the adult pig proximal humerus. The unfused epiphyses of the juvenile pig are also shown.

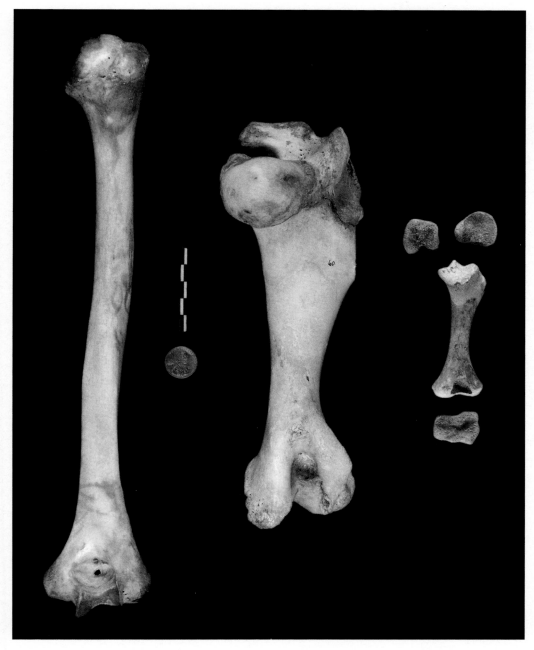

Fig. 6-02. A human right humerus (posterior view) is compared to adult and juvenile pig right humeri (caudal views). The unfused epiphyses of the juvenile pig are also shown.

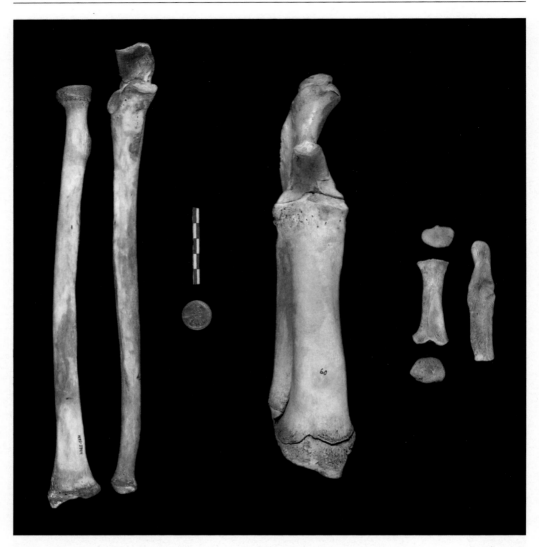

Fig. 6-03. A right human radius and ulna (anterior views) is compared to adult and juvenile right pig radii and ulnae (cranial views). In the adult pig, the radius is attached to the ulna. The unfused epiphyses of the juvenile pig radius are also shown.

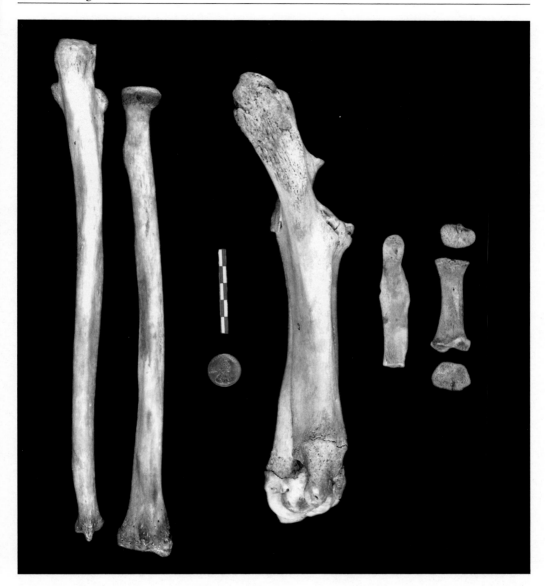

Fig. 6-04. A right human radius and ulna (posterior views) are compared to adult and juvenile right pig radii and ulnae (caudal views). The unfused epiphyses of the juvenile pig radius are also shown.

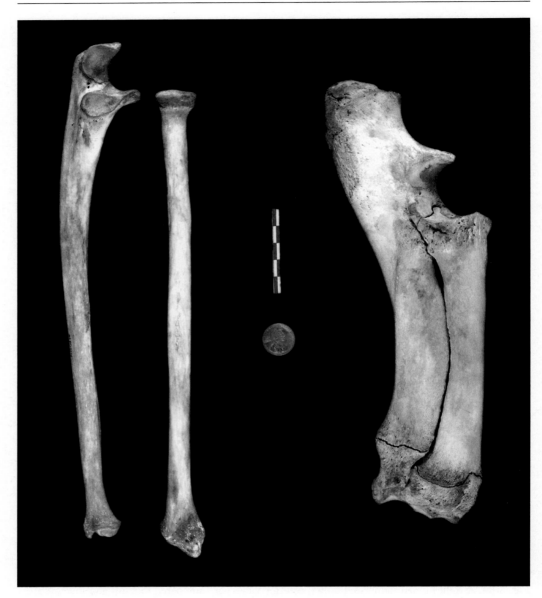

Fig. 6-05. A right human radius and ulna (lateral views) are compared to an adult right pig radius and ulna (lateral view). Note the large size of the pig's olecranon process.

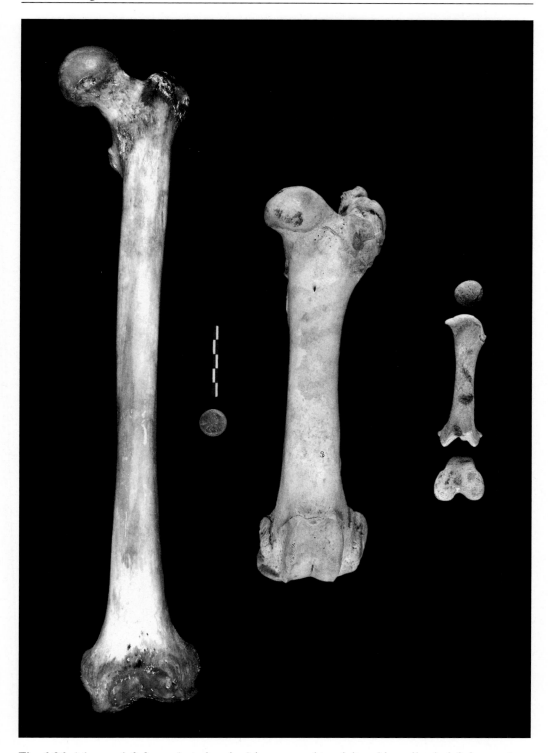

Fig. 6-06. A human left femur (anterior view) is compared to adult and juvenile pig left femora (cranial views). Note the larger and more developed greater trochanter on the adult pig's femur. The unfused epiphyses of the juvenile pig are also shown.

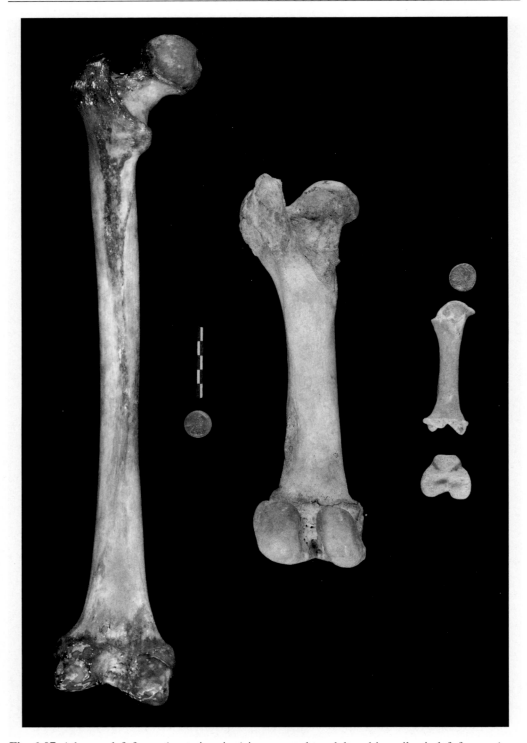

Fig. 6-07. A human left femur (posterior view) is compared to adult and juvenile pig left femora (caudal views). Note the presence of the linea aspera on the human femur. The unfused epiphyses of the juvenile pig are also shown.

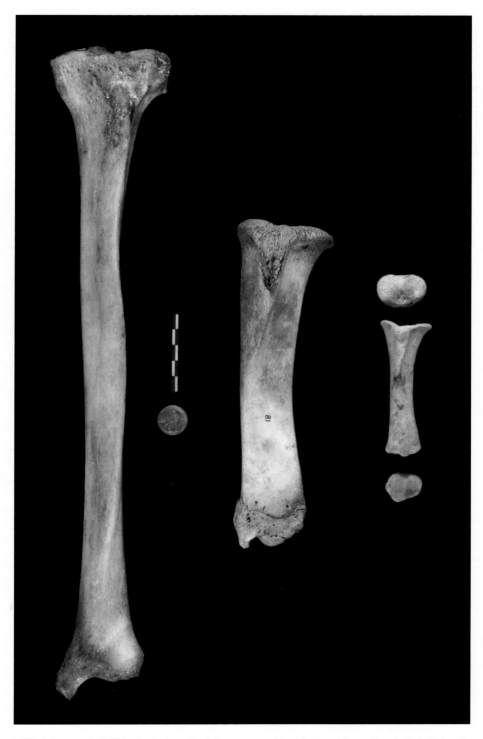

Fig. 6-08. A human left tibia (anterior view) is compared to adult and juvenile pig left tibiae (cranial views). The distal pig tibia has two parallel concave articular facets for articulation with the astragalus. Note that the proximal epiphysis of the large pig is unfused. The unfused epiphyses of the juvenile pig are also shown.

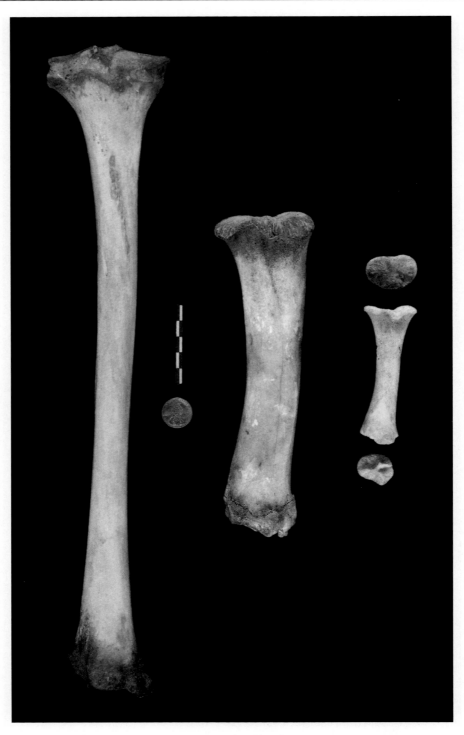

Fig. 6-09. A human left tibia (posterior view) is compared to adult and juvenile pig left tibiae (caudal views). Note that the proximal epiphysis of the large pig is unfused. The unfused epiphyses of the juvenile pig are also shown.

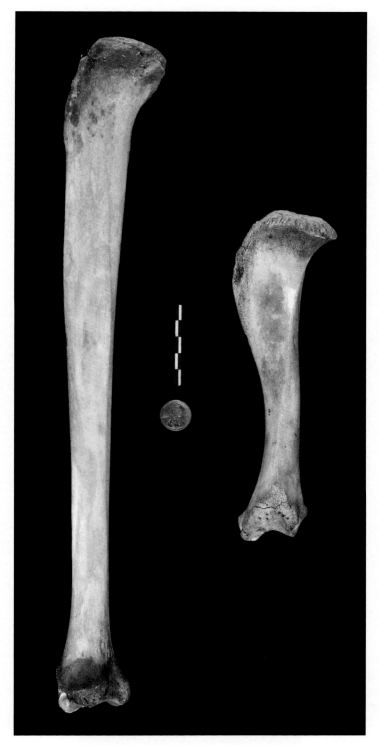

Fig. 6-10. A human left tibia (lateral view) is compared to an adult pig left tibia (lateral view). Note that the proximal epiphysis of the large pig is unfused.

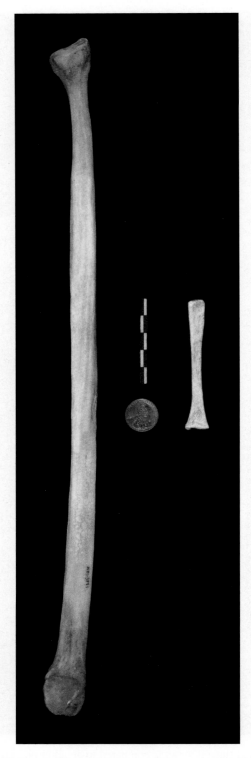

Fig. 6-11. A human right fibula (medial view) is compared to a juvenile pig right fibula (medial view).

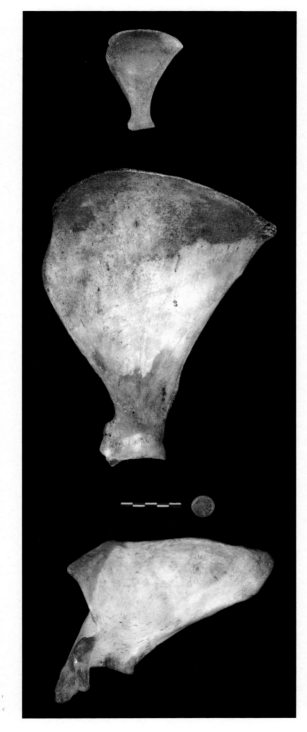

Fig. 6-12. A human right scapula (anterior view) is compared to adult and juvenile pig right scapulae (medial views). Note that the large human acromion process is not entirely blocked by the body of the scapula and is visible in the anterior view.

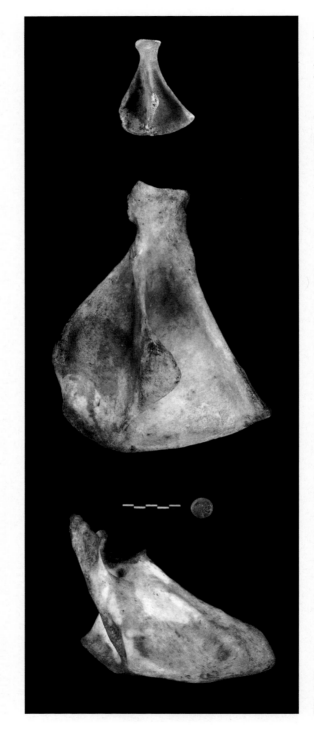

Fig. 6-13. A human right scapula (posterior view) is compared to adult and juvenile pig right scapulae (lateral views). The pig scapula has a large tuberosity of the spine but only a rudimentary acromion process.

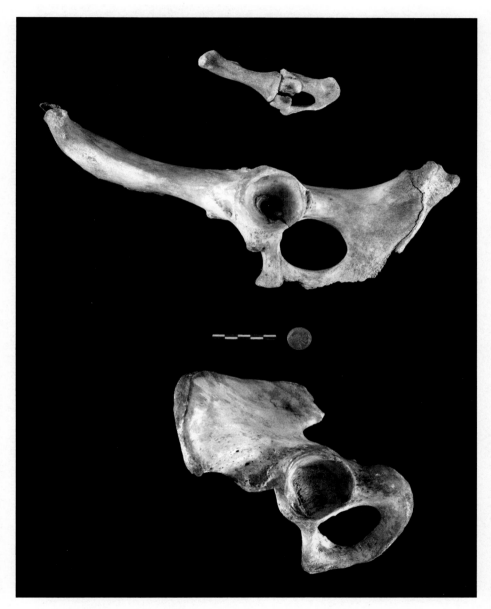

Fig. 6-14. A human left innominate (lateral view) is compared to adult and juvenile pig left innominates (lateral views). Note the large size of the ischial tuberosity on the adult pig's pelvis.

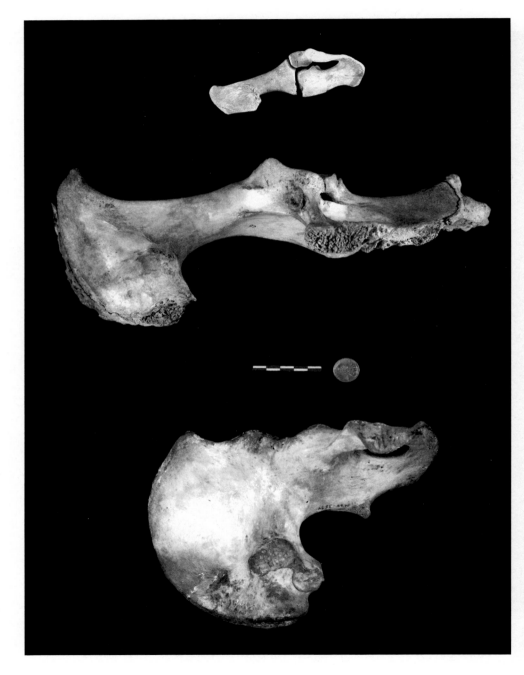

Fig. 6-15. A human left innominate (medial view) is compared to adult and juvenile pig left innominates (medial views). Gaps in fusion of the pelvic bones are visible on the juvenile pig.

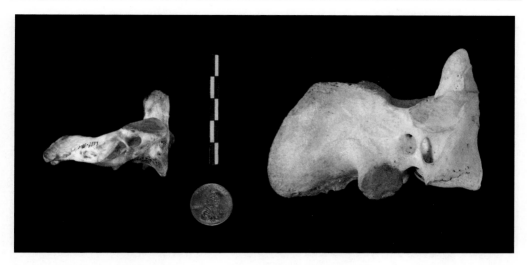

Fig. 6-16. Human axis (C2) is compared with an adult pig axis (C2). Both views are lateral.

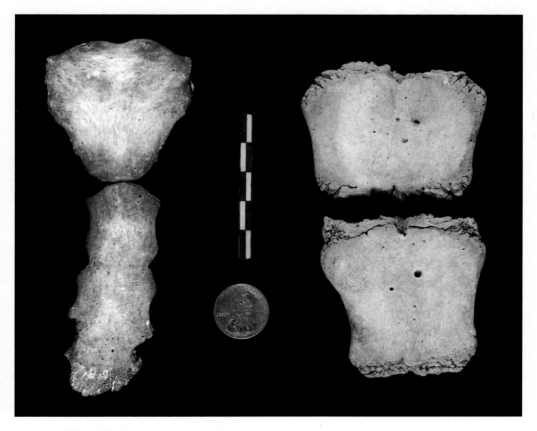

Fig. 6-17. A human sternum (anterior view) is compared to two pig sternebrae.

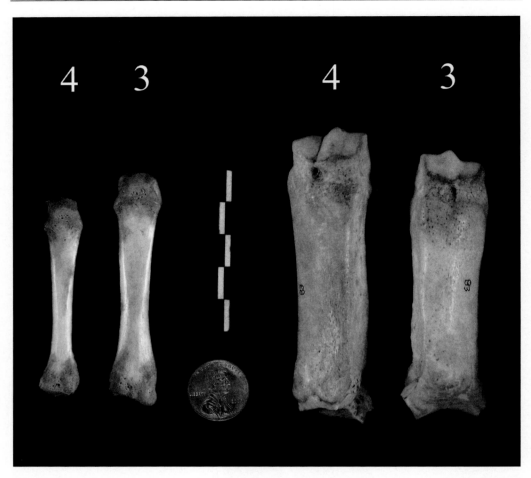

Fig. 6-18. Human 3rd and 4th left metacarpals (anterior views) are compared to adult pig left 3rd and 4th metacarpals (dorsal views). The pig is an artiodactyl or even-toed ungulate. It has four metapodia on each foot. The 3rd and 4th metapodia are larger central metapodia, while 2nd and 5th metapodia are smaller.

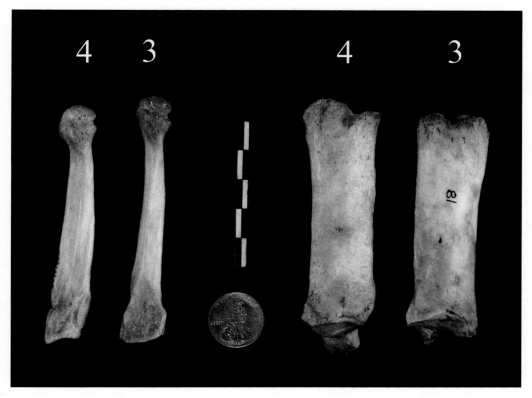

Fig. 6-19. Human 3rd and 4th left metatarsals (superior views) are compared to adult pig left 3rd and 4th metatarsals (dorsal views). Note that distal pig metatarsals are unfused. The pig is an artiodactyl or even-toed ungulate. It has four metapodia on each foot. The 3rd and 4th metapodia are larger central metapodia, while 2nd and 5th metapodia are smaller.

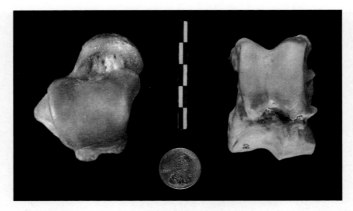

Fig. 6-20. Human left talus (superior view) is compared to an adult pig left astragalus (dorsal view). Note that the pig astragalus has the "double-pulley" form that is characteristic of the artiodactyls.

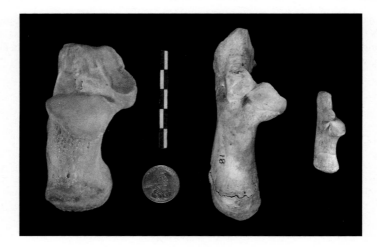

Fig. 6-21. Human left calcaneus (superior view) is compared to adult and juvenile pig left calcanei (dorsal views).

7   Human vs Goat

Fig. 7-00. Goat skull lateral view. The adult goat's dental formula is 0/3.0/1.3/3.3/3. The juvenile formula is 0/3.0/1.3/3.

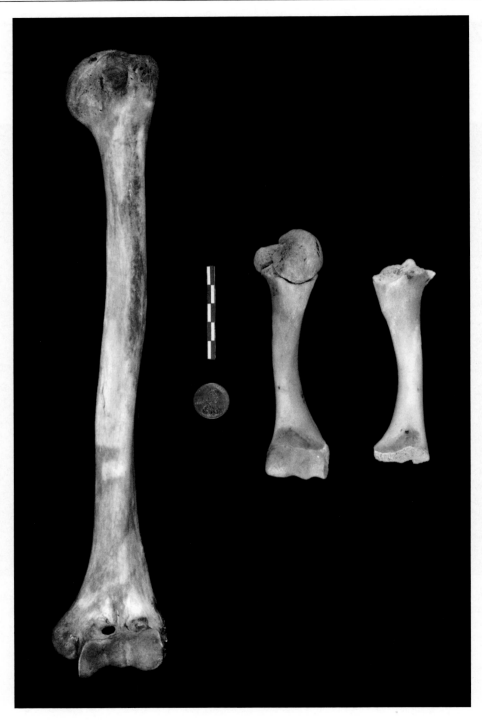

Fig. 7-01. Human left humerus (anterior view) is compared to a goat's left humerus (cranial view with epiphyses) and a goat's right humerus (cranial view, without epiphyses).

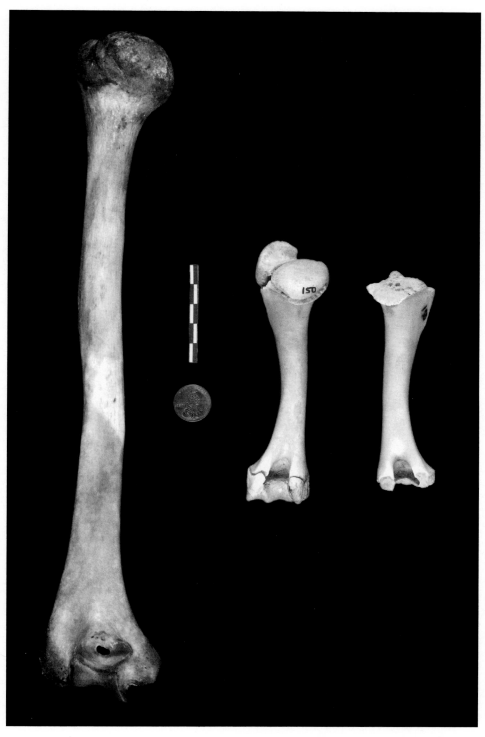

Fig. 7-02. Human left humerus (posterior view) is compared to a goat's left humerus (caudal view with epiphyses) and a goat's right humerus (caudal view, without epiphyses).

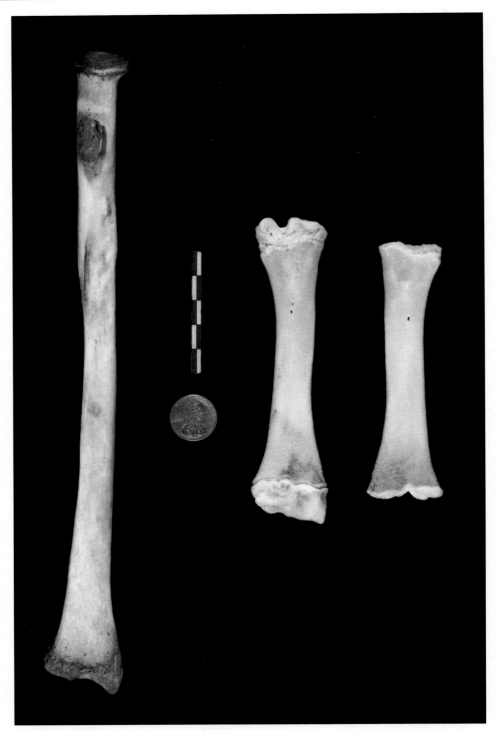

Fig. 7-03. Human left radius (anterior view) is compared to a goat's left radius (caudal view, with epiphyses) and a goat's right radius (caudal view, without epiphyses).

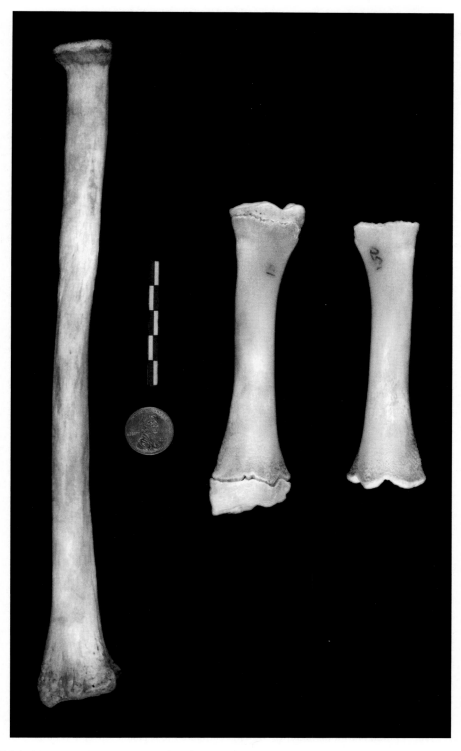

Fig. 7-04. A human left radius (posterior view) is compared to a goat's left radius (cranial view, with epiphyses) and a goat's right radius (cranial view, without epiphyses).

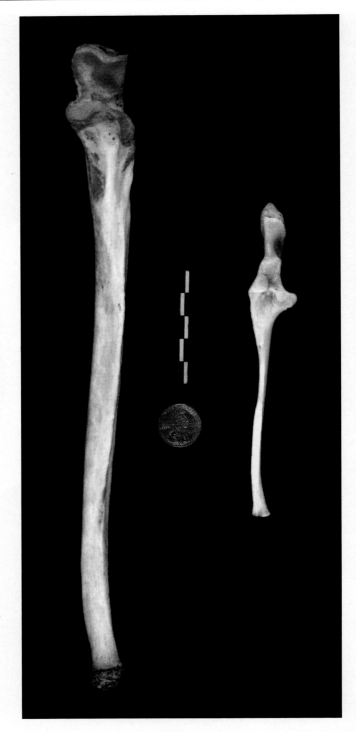

Fig. 7-05. A human left ulna (anterior view) is compared to a goat's left ulna (cranial view).

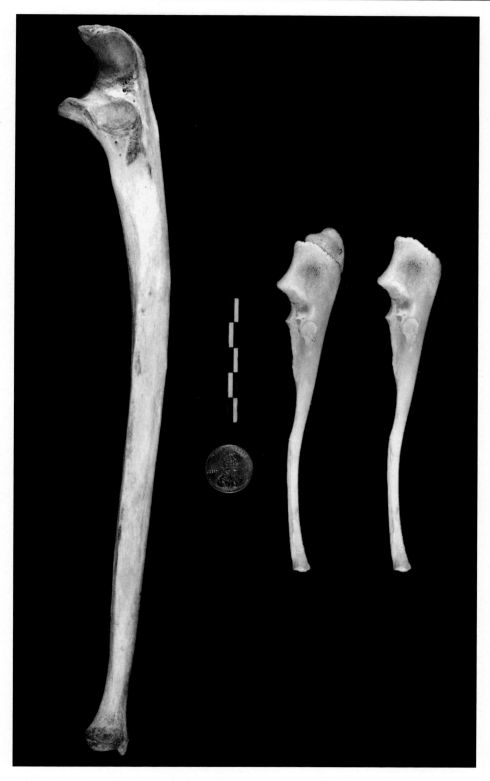

Fig. 7-06. A human left ulna (lateral view) is compared to a goat's left ulna (lateral views, with and without epiphyses).

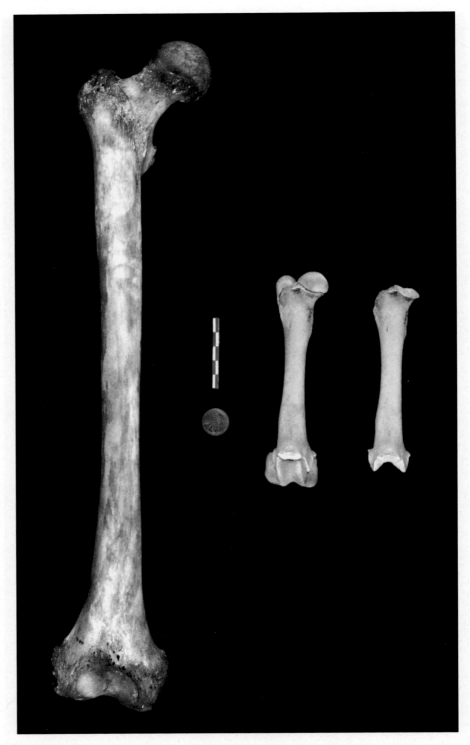

Fig. 7-07. Human right femur (anterior view) is compared to a goat's right femur (cranial view, with and without epiphyses).

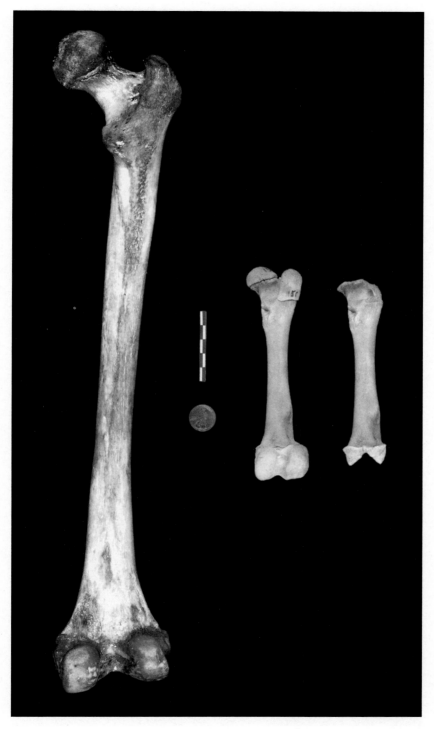

Fig. 7-08. Human right femur (posterior view) is compared to a goat's left femur (caudal view, with and without epiphyses).

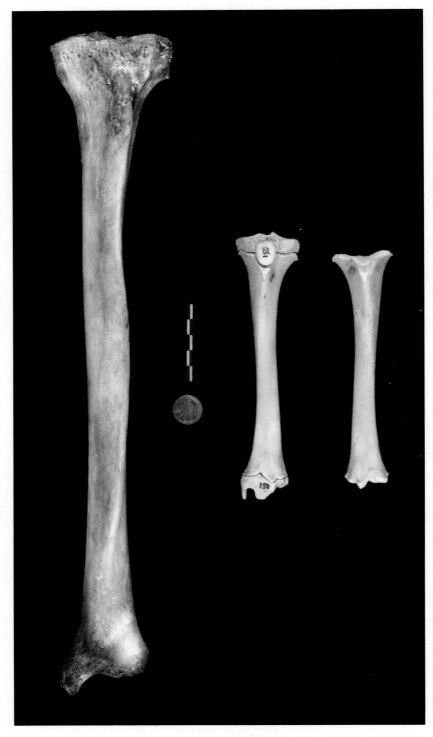

Fig. 7-09. A human left tibia (anterior view) is compared to a goat's left tibia (cranial view, with epiphyses) and a goat's right tibia (cranial view, without epiphyses).

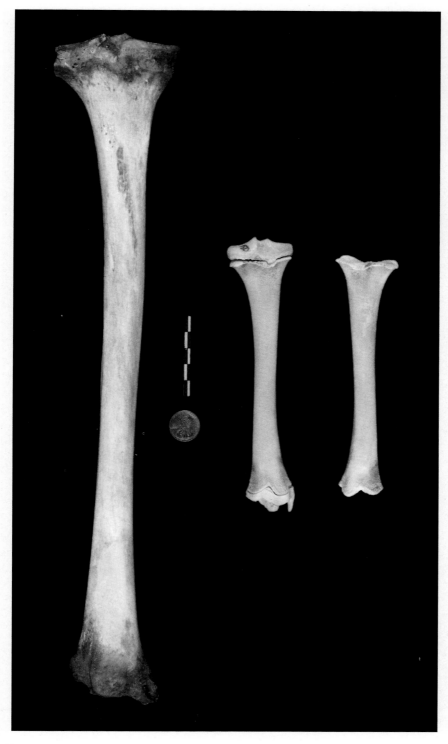

Fig. 7-10. A human left tibia (posterior view) is compared to a goat's left tibia (caudal view, with epi-physes) and a goat's right tibia (caudal view, without epiphyses).

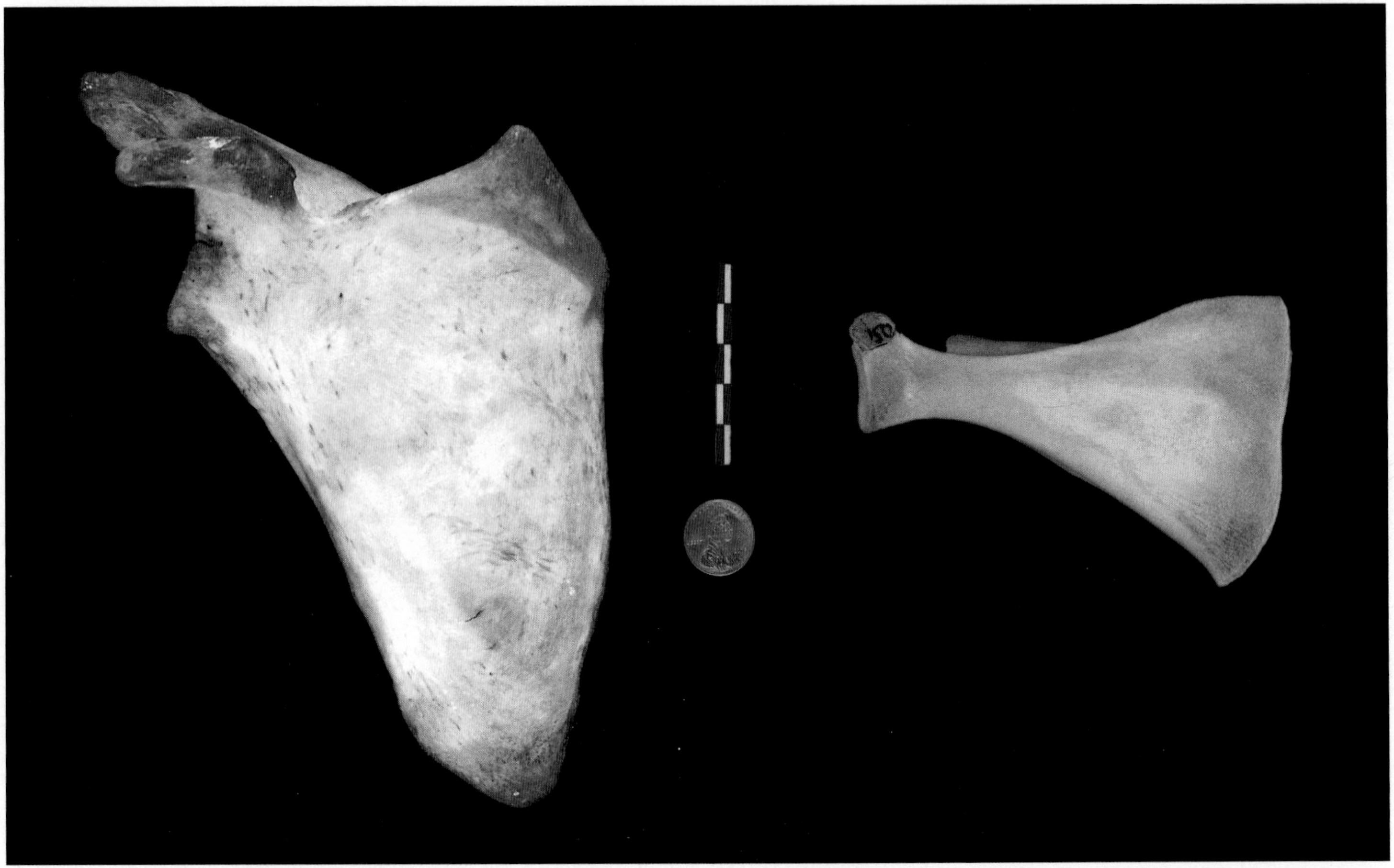

Fig. 7-11. Human right scapula (anterior view) compared to a goat's right scapula (medial view).

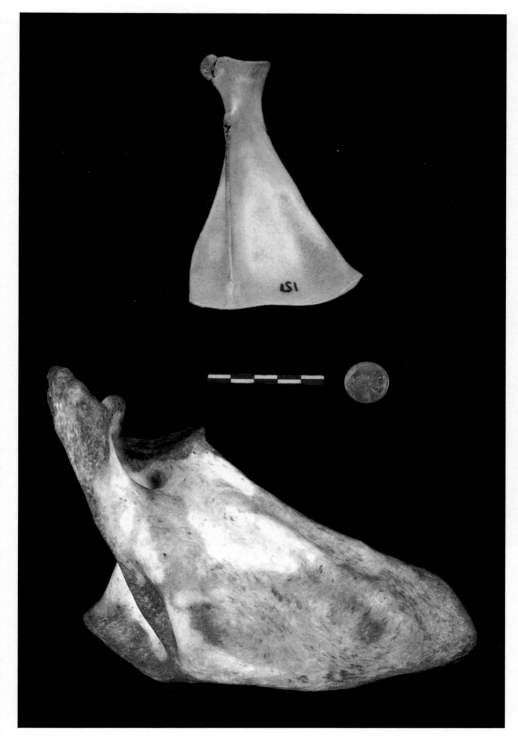

Fig. 7-12. Human right scapula (posterior view) is compared to the goat's right scapula (lateral view). Note the large acromion process on the human scapula.

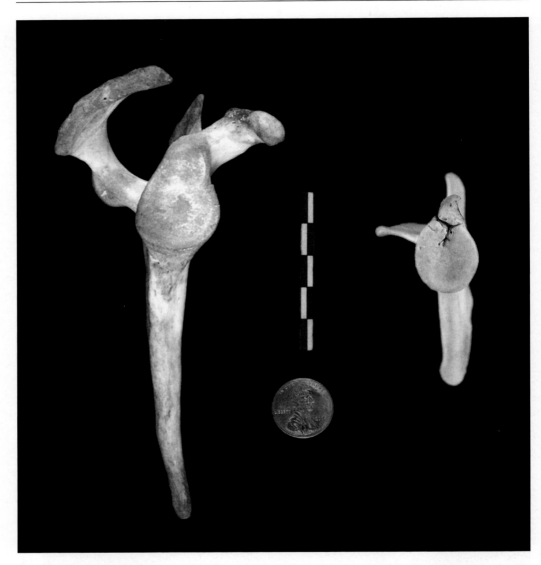

Fig. 7-13. Human right scapula (lateral view) is compared to a goat's right scapula (distal view). The goat's and sheep's glenoid cavity is generally more oval in shape, while the deer's is more rounded.

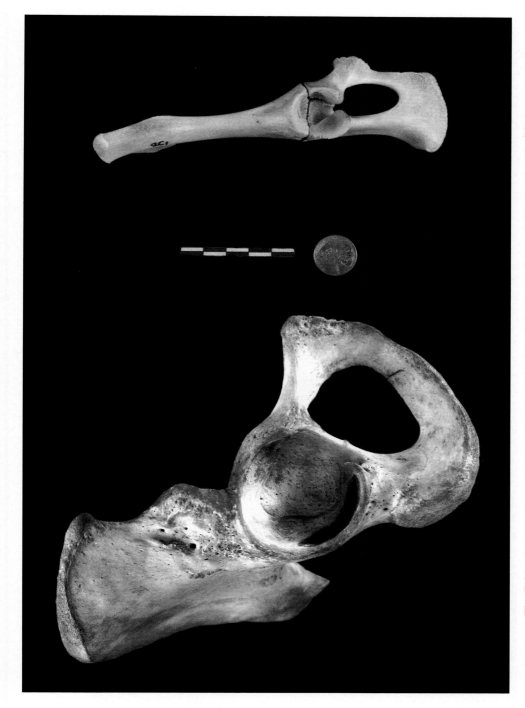

Fig. 7-14. A human right innominate (lateral view) is compared to a goat's right innominate (lateral view).

131

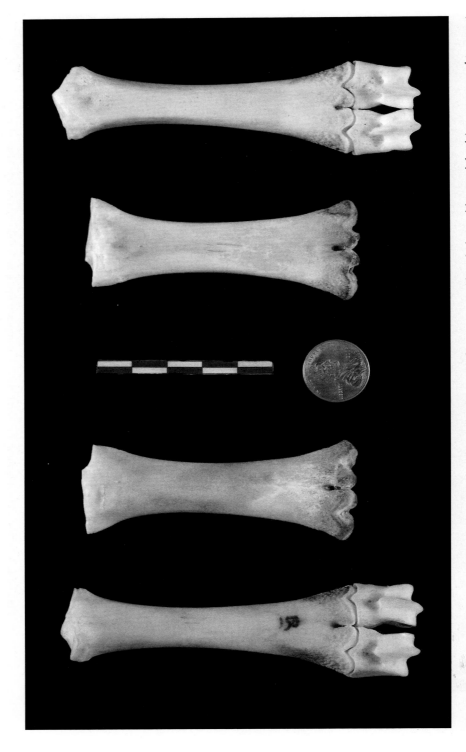

Fig. 7-15. A goat left metatarsus (dorsal view, with epiphysis) and a goat left metacarpus (dorsal view, without epiphysis) are compared to a goat left metatarsus (plantar view, with epiphysis) and a goat left metacarpus (volar view, without epiphysis). The goat metacarpus and metatarsus are composed of the fused 3rd and 4th metacarpals and metatarsals.

8   Human vs Sheep

Fig. 8-00. Sheep's cranium and mandible (lateral view). The sheep has no upper incisors and canines. Each maxilla includes three premolars and three molars. Each mandible includes three incisors and an incisiform canine, followed by a long diastema, three premolars and three molars.

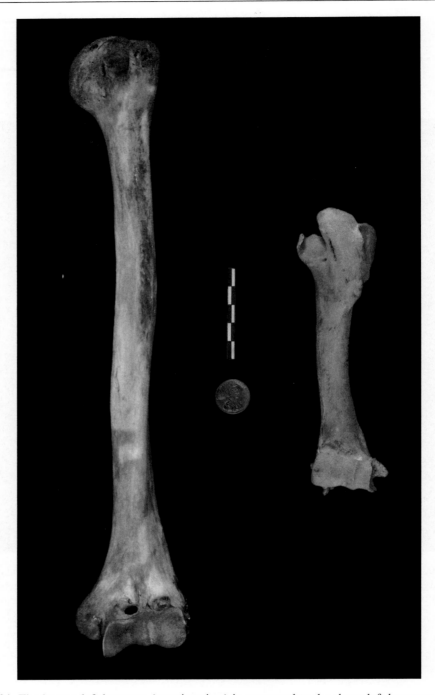

Fig. 8-01. The human left humerus (anterior view) is compared to the sheep left humerus (cranial view). A slight exostosis is apparent on the lateral epicondyle of the sheep's distal humerus. This is an example of "penning elbow," a pathological condition that is relatively common in domestic sheep and is caused by trauma to the joint area. Note also that the greater tubercle is more well-developed in the sheep's humerus than it is in human humerus.

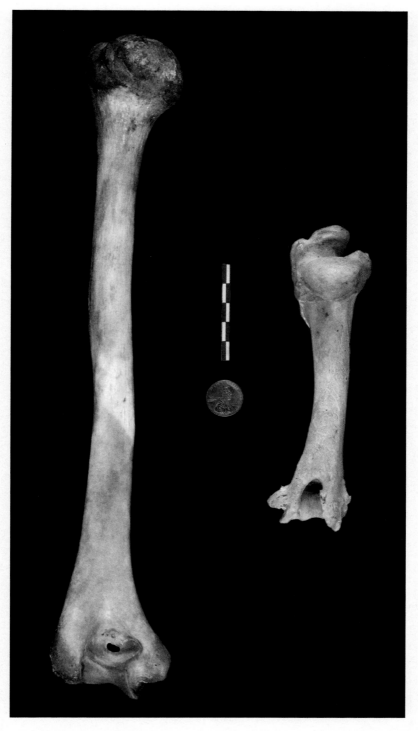

Fig. 8-02. A human left humerus (posterior view) is compared to a sheep's left humerus (caudal view). Note that the overall morphology of the sheep's skeleton is quite similar to the cow's, but that the sheep is significantly smaller.

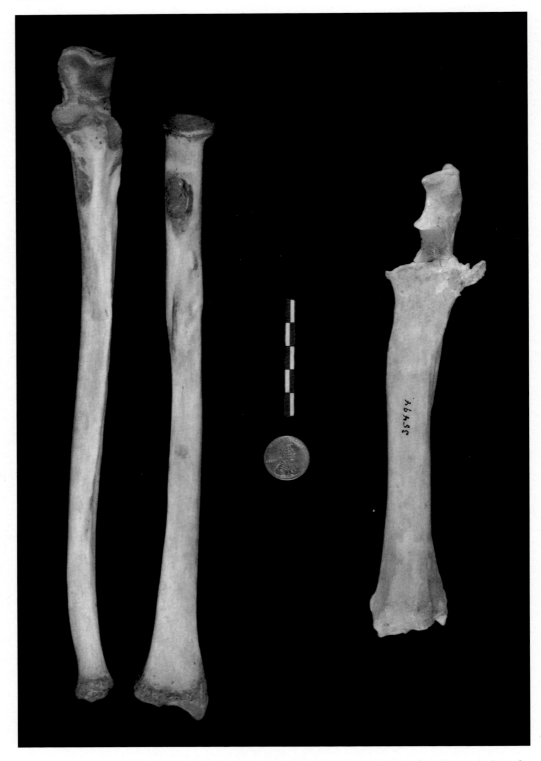

Fig. 8-03. A human left ulna and radius (anterior views) are compared to a left radius and ulna of a sheep (cranial view). Note that the sheep's radius is large in relation to the ulna. In most adult sheep, the shaft of the radius is fused to the ulna.

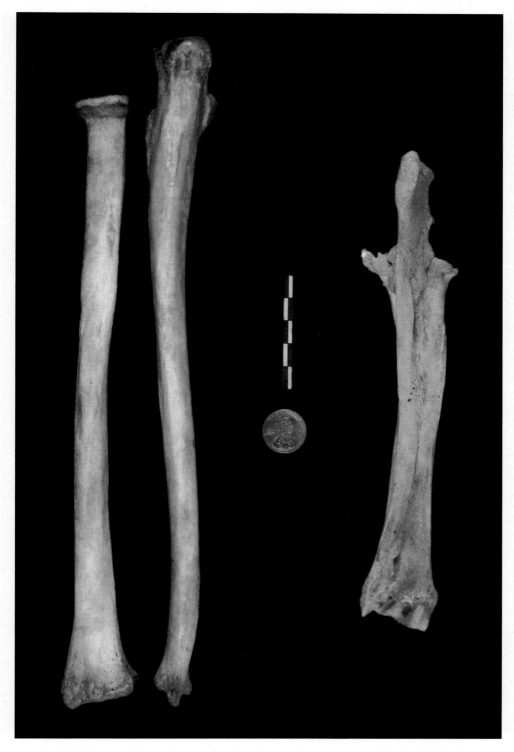

Fig. 8-04. A left human radius and ulna (posterior views) are compared to a sheep's left radius and ulna (caudal view). Note that the shaft of the sheep's ulna is fused to the caudal surface of the radius. The exostosis associated with "penning elbow" can also be seen on the lateral side of the proximal radius.

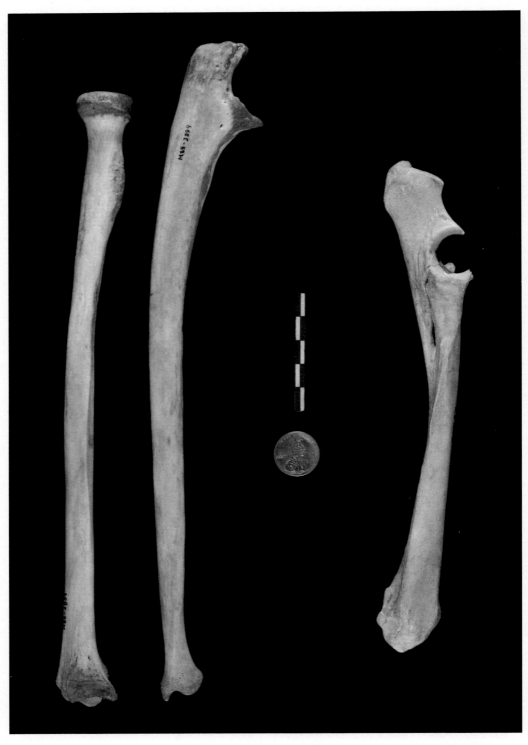

Fig. 8-05. A human left radius and ulna (medial views) are compared to a sheep's left radius and ulna (medial view). Although the shaft of the sheep's ulna is very slender, the olecranon process is well-developed when compared to the human olecranon process.

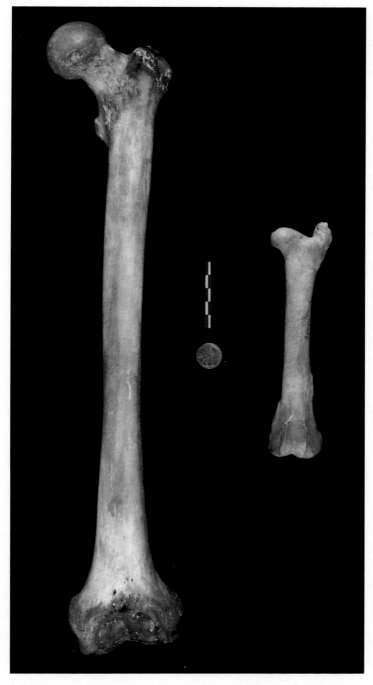

Fig. 8-06. The human left femur (anterior view) is compared to the sheep's left femur (cranial view). The most proximal point of the human femur is the head, while the most proximal point on the sheep's femur is the greater trochanter. Sheep and goat are quite similar (both are members of the tribe Caprini), and the differences between their postcranial skeletons are quite subtle. The angle formed between the head and the greater trochanter is generally obtuse in sheep, while it is closer to a right angle in goats.

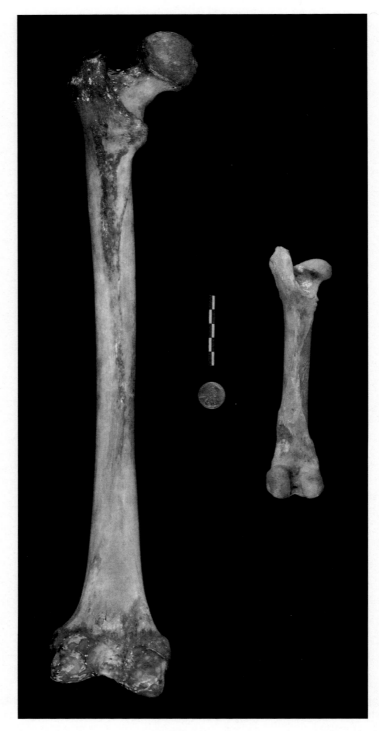

Fig. 8-07. The human left femur (posterior view) is compared to the sheep's left femur (caudal view). The sheep femur is marked by a supercondylar fossa on the lateral portion of the shaft. The linea aspera is a distinctively human feature.

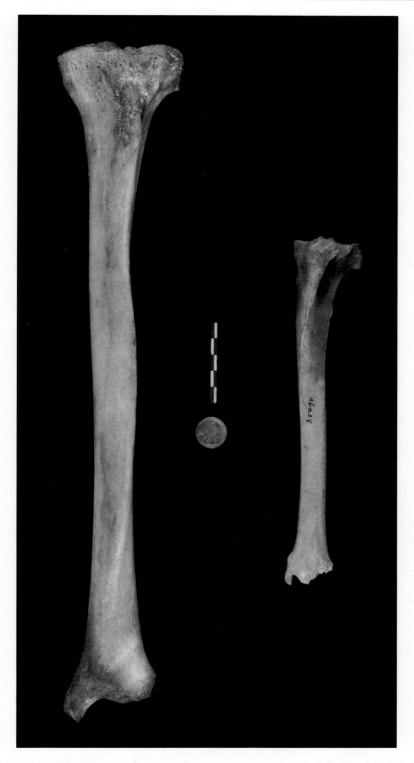

Fig. 8-08. A human left tibia (anterior view) is compared to a sheep's left tibia (cranial view). The distal end of the sheep's tibia has two parallel articular facets for articulation with the astragalus.

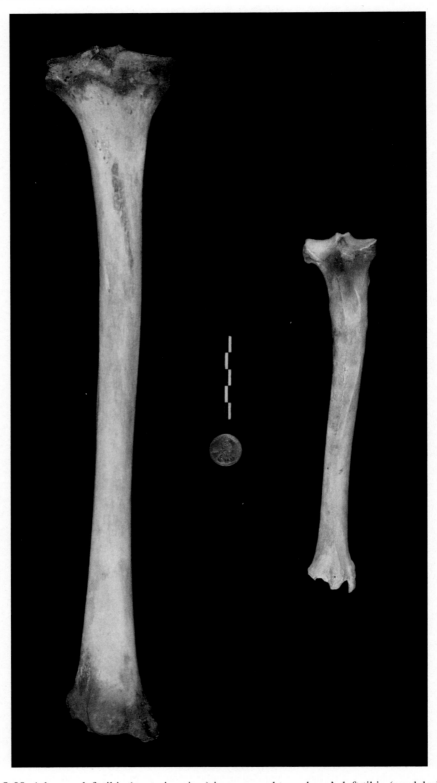

Fig. 8-09. A human left tibia (posterior view) is compared to a sheep's left tibia (caudal view).

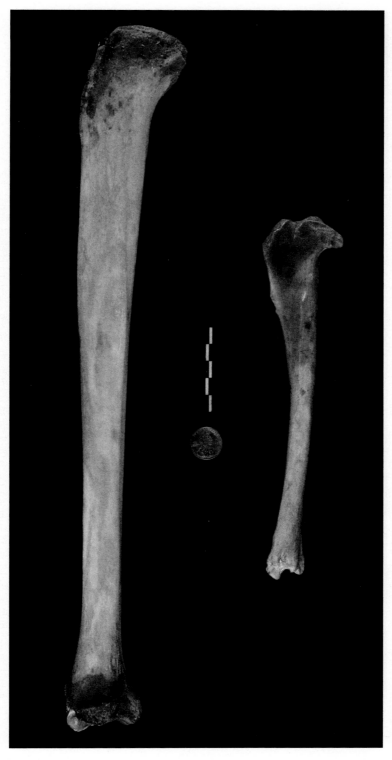

Fig. 8-10. A human left tibia (medial view) is compared to a sheep's left tibia (medial view). Note that the tibial tuberosity and crest are well developed in the sheep.

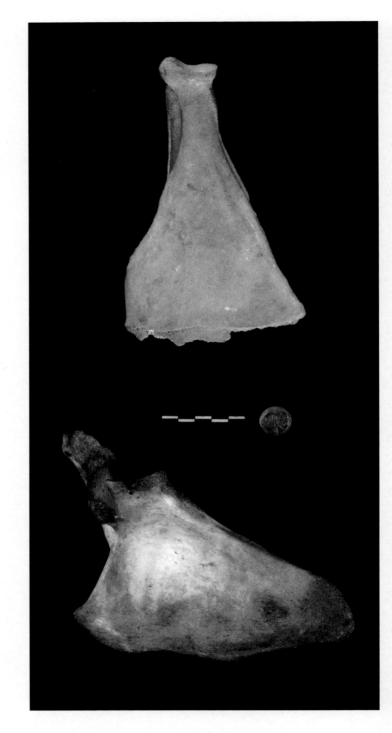

Fig. 8-11. A human left scapula (anterior view) is compared with a sheep's left scapula (medial view). Both scapulae are oriented as they would be in a human skeleton. In comparison to the human scapula, the sheep's scapula is elongated.

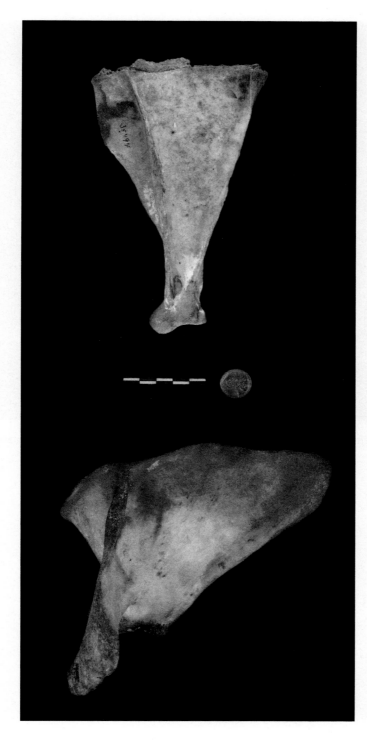

Fig. 8-12. A human left scapula (posterior view) is compared to a sheep's left scapula (lateral view). The acromion process in the sheep is quite small, while it is quite large in the human. In addition, a small crest is visible on the caudal border of the neck of the sheep's scapula. The presence of this crest can be used to distinguish sheep from goats.

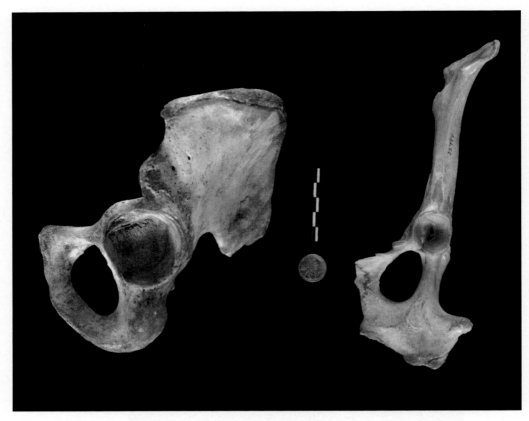

Fig. 8-13. A human left innominate (lateral view) is compared to a sheep's left innominate (lateral view).

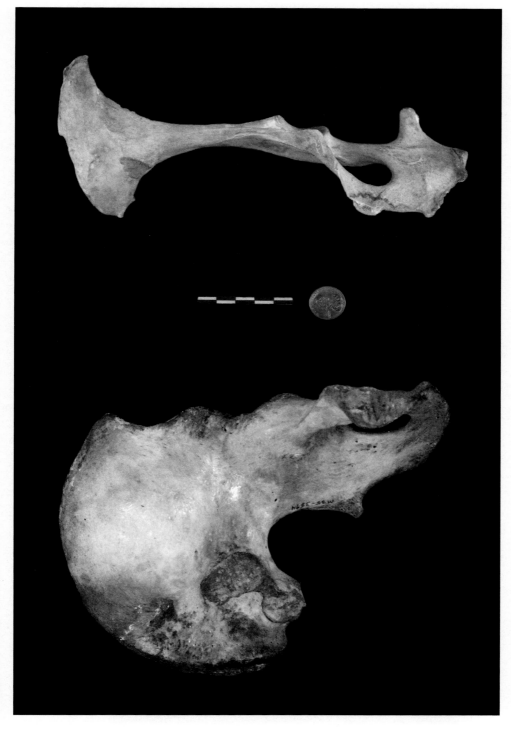

Fig. 8-14. A human left innominate (medial *view*) is compared to a sheep's left innominate (ventral *view*). Note the sheep's elongated ilium.

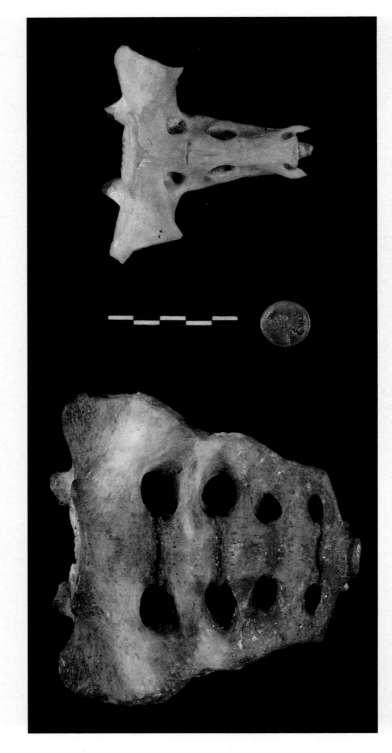

Fig. 8-15. The human sacrum (anterior view) is compared to the sheep's sacrum (ventral view). The wings of the sheep's sacrum are quite wide, while the rest of the sacrum is relatively narrow.

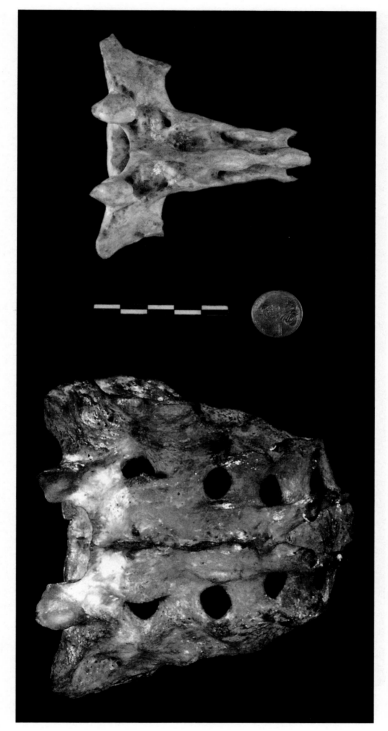

Fig. 8-16. The human sacrum (posterior view) is compared to the sheep's sacrum (dorsal view).

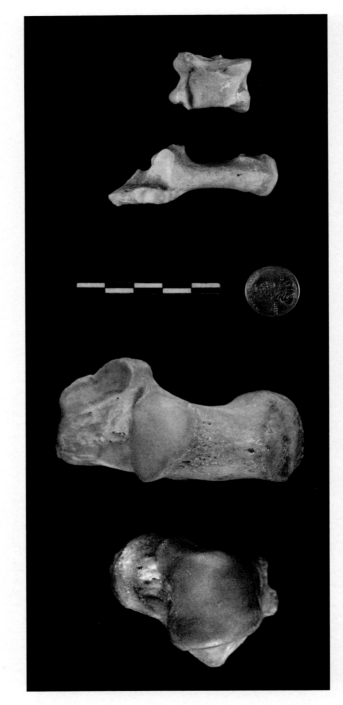

Fig. 8-17. Human left talus and calcaneus (superior views) are compared to a sheep's left calcaneus (dorsal view) and a sheep's left astragalus (plantar view). The sheep's astragalus has the "double pulley" form that is seen in all the artiodactyls. The sheep's calcaneus is elongated and includes a dorsal articular facet for the malleolus (the evolutionary remnant of the distal fibula).

150

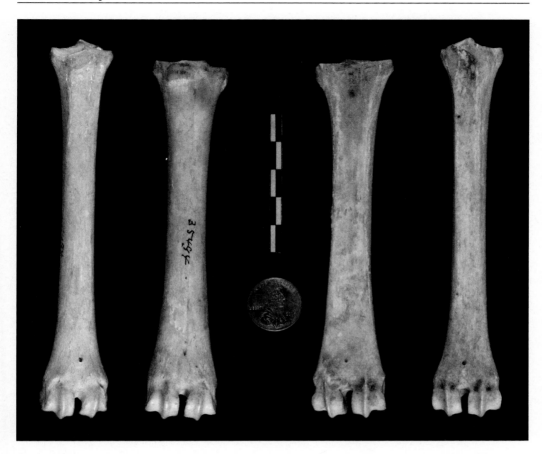

Fig. 8-18. Sheep left metacarpus and metatarsus (dorsal views) are shown on the left. The sheep metacarpus (palmar view) and the sheep metatarsus (plantar view) are shown on the right. The sheep metapodia are composed of the fused third and fourth metacarpals and metatarsals.

9   Human vs Dog

Fig. 9-00. Lateral view of the dog cranium and mandible. The dental formula for the adult dog includes 3 incisors, 1 large canine, 4 premolars, and two molars on the maxilla, and 3 incisors, 1 canine, 4 premolars, and 3 molars on the mandible.

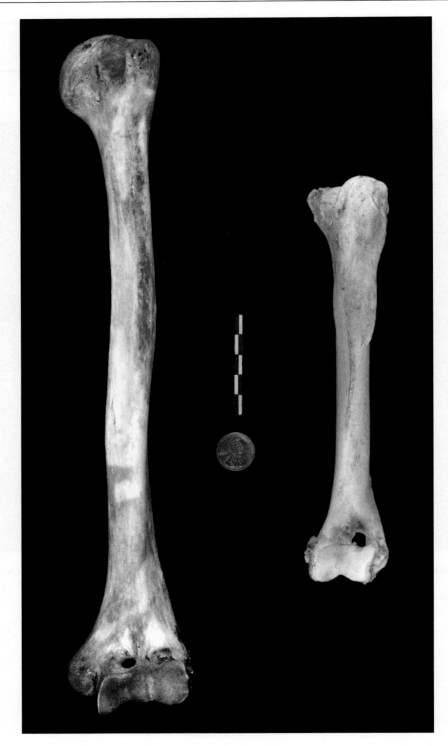

Fig. 9-01. A human left humerus (anterior view) is compared to a dog's left humerus (cranial view). One of the features of the dog's humerus is that the radial fossa communicates with the olecranon fossa, producing a supratrochlear foramen. These foramina (also called septal apertures) are sometimes seen in humans.

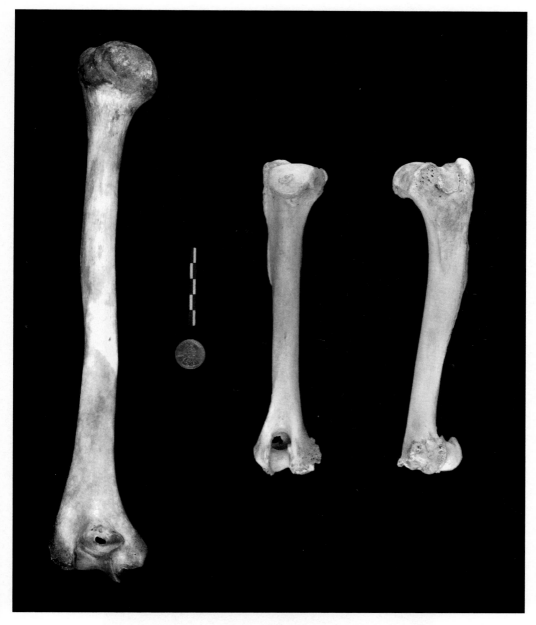

Fig. 9-02. A human left humerus (posterior view) is compared to a dog left humerus (caudal and medial views). When compared to the human head of the humerus, the head of the humerus in the dog is elongated sagittally.

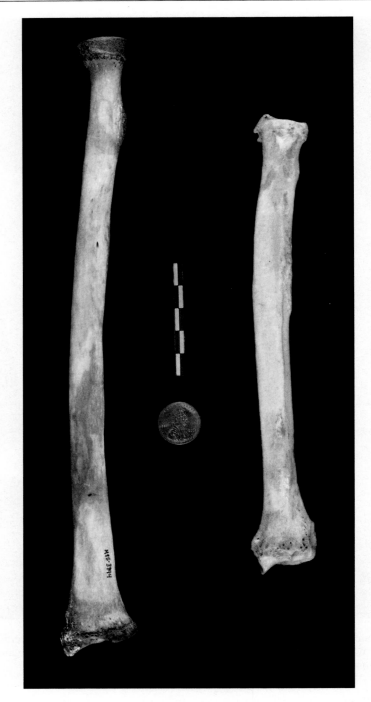

Fig. 9-03. A human right radius (anterior view) compared to a dog's right radius (caudal view). Human skeletons are oriented with the palms forward, so that the radius and ulna are not crossed. Quadrupedal animals are oriented with their paws facing the ground. This means that the proximal ulna is medial to the radius, while the distal ulna is on the lateral side. Both the human and the dog's radius have a distinctive head and neck.

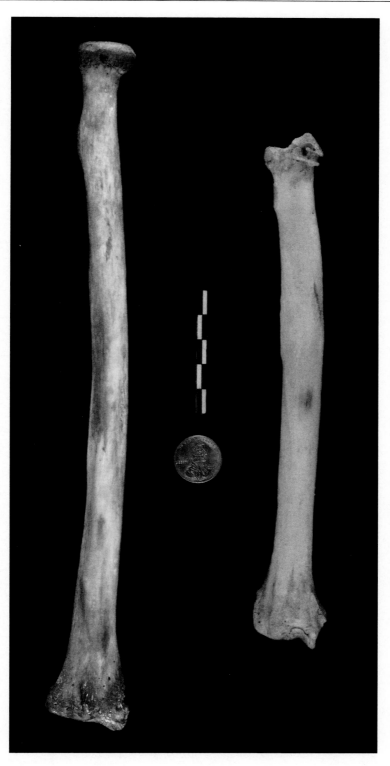

Fig. 9-04. A human right radius (posterior view) is compared to a dog's right radius (cranial view). Note the elongated neck on the human radius.

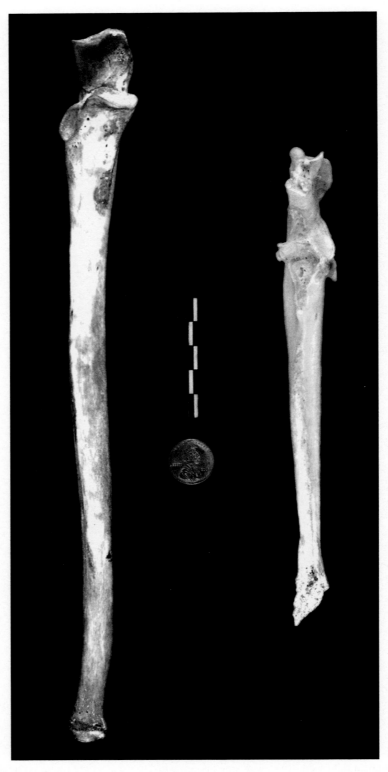

Fig. 9-05. A human right ulna (anterior view) is compared to a dog's right ulna (cranial view). The distal portion of the ulna (styloid process) is missing on this dog, who was an elderly and arthritic individual.

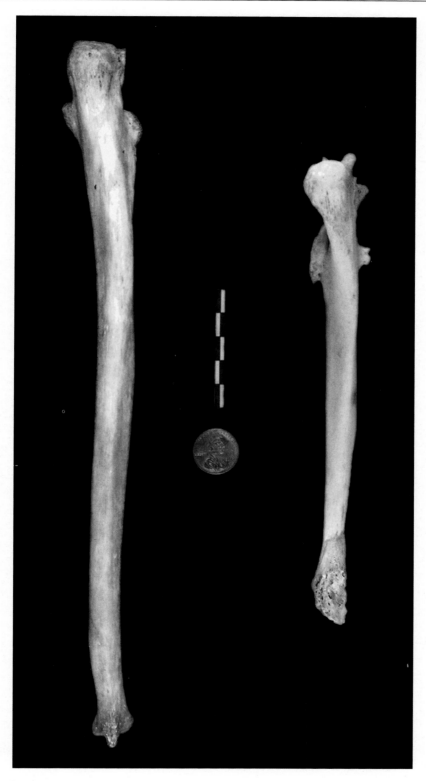

Fig. 9-06. A human right ulna (posterior view) is compared to a dog's right ulna (caudal view).

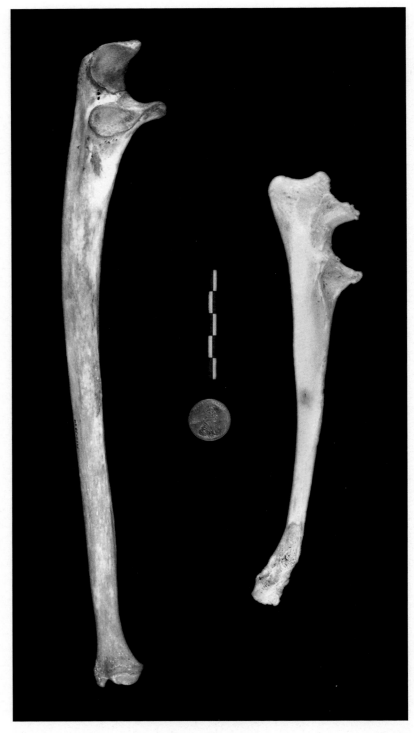

Fig. 9-07. A human right ulna (medial view) is compared to a dog's right ulna (medial view). The olecranon is larger and more well-developed on the dog.

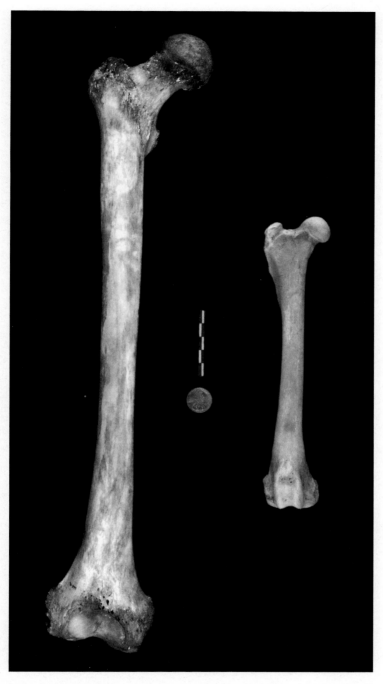

Fig. 9-08. A human right femur (anterior view) is compared to a dog's right femur (cranial view). When compared to the human femur, the dog's greater trochanter is more well developed. The human femur shows a "kneeing in" that is not seen in the dog femur, since in humans the hips are widely spaced, but the knees are much closer to the center of the body. This orientation of the leg under the torso in humans is also referred to as the valgus knee.

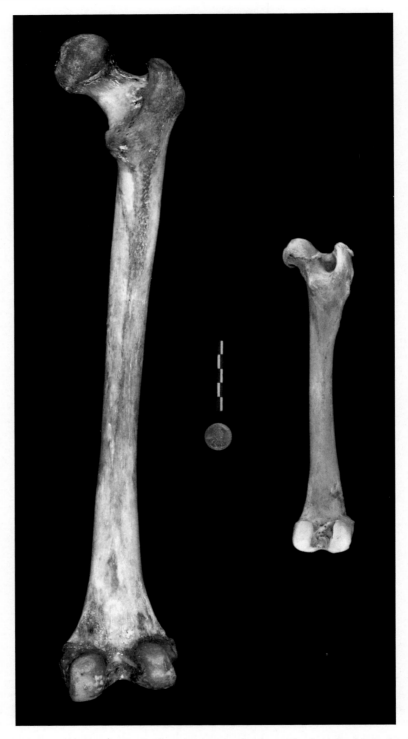

Fig. 9-09. A human right femur (posterior view) compared to a dog's right femur (caudal view). Two small articular facets are visible on the dog's femur, just proximal to the distal condyles. These are the medial and lateral articular facets for two sesamoids that are located in the origin of the gastrocnemius muscle.

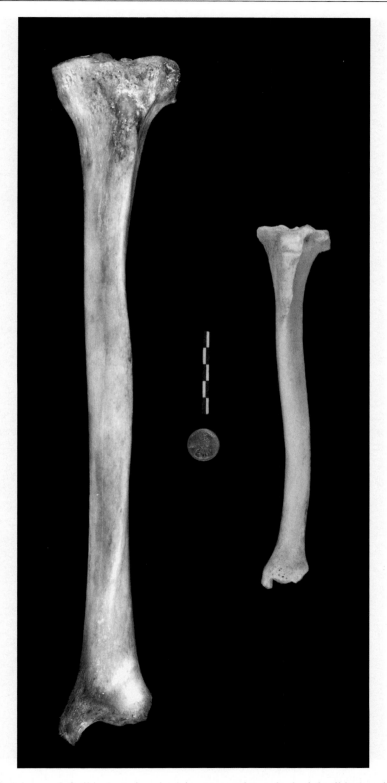

Fig. 9-10. A human left tibia (anterior view) is compared to a dog's right tibia (cranial view).

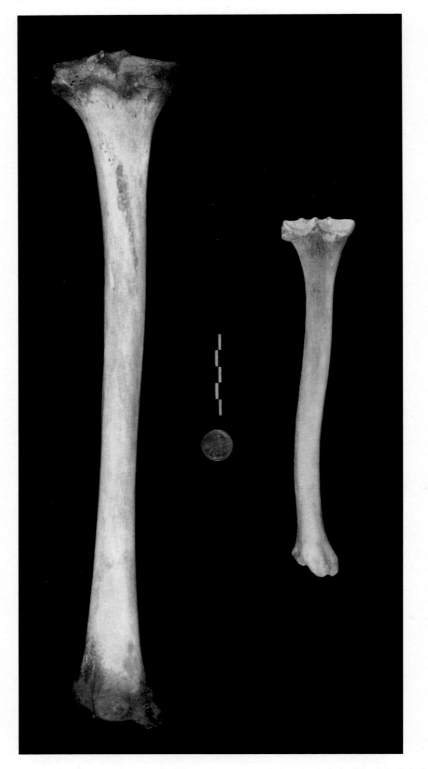

Fig. 9-11. A human left tibia (posterior view) is compared to a dog's left tibia (caudal view).

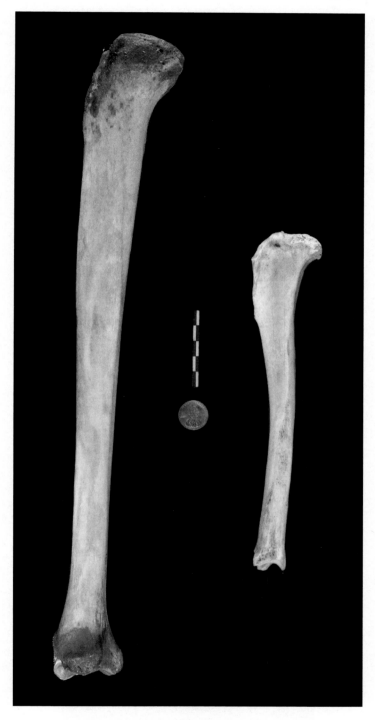

Fig. 9-12. A human left tibia (lateral view) is compared to a dog's left tibia (lateral view). Although these two tibias are broadly similar in form, the proximal tibia of the dog has a more well-developed tibial tuberosity.

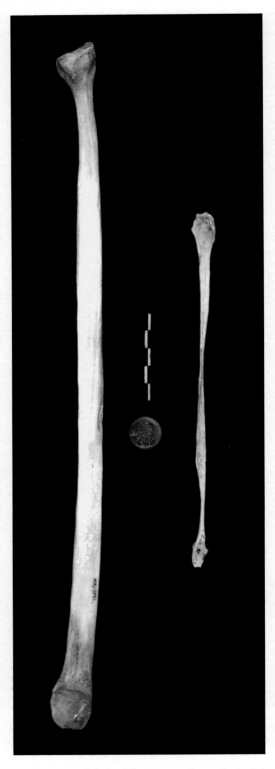

Fig. 9-13. A human right fibula (medial view) is compared to a dog's right fibula (medial view).

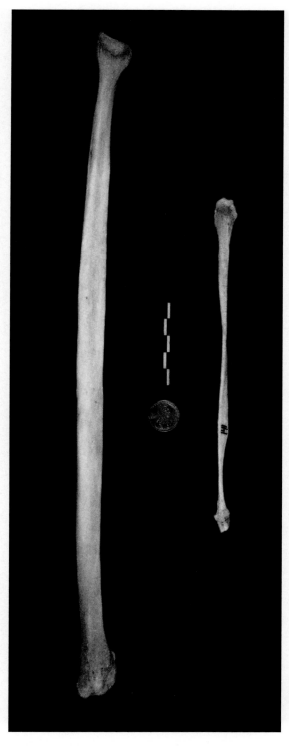

Fig. 9-14. A human right fibula (lateral view) is compared to a dog's right fibula (lateral view). The body of the dog's fibula is quite slender.

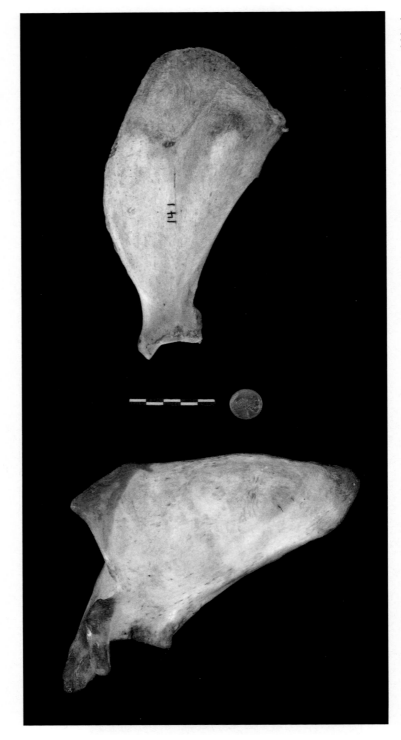

Fig. 9-15. A human right scapula (anterior view) is compared to a dog's right scapula (medial view). Both bones are oriented as they would be in a human. Note that the dog's scapula is elongated.

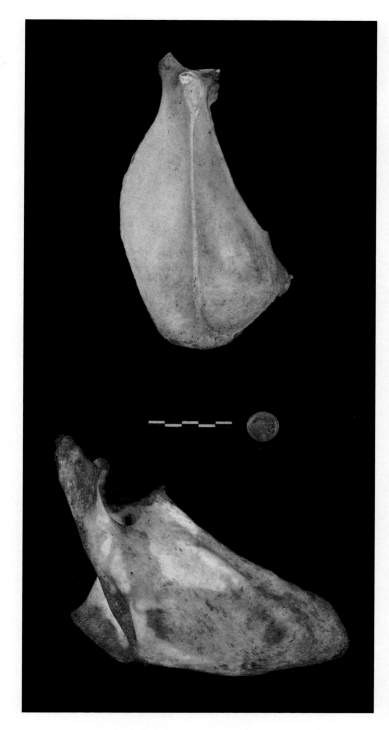

Fig. 9-16. A human right scapula (posterior view) is compared to a dog's right scapula (lateral view). Note that the spine of the dog's scapula divides the scapula into two nearly equal halves. The human scapula's larger acromion process is visible in this view.

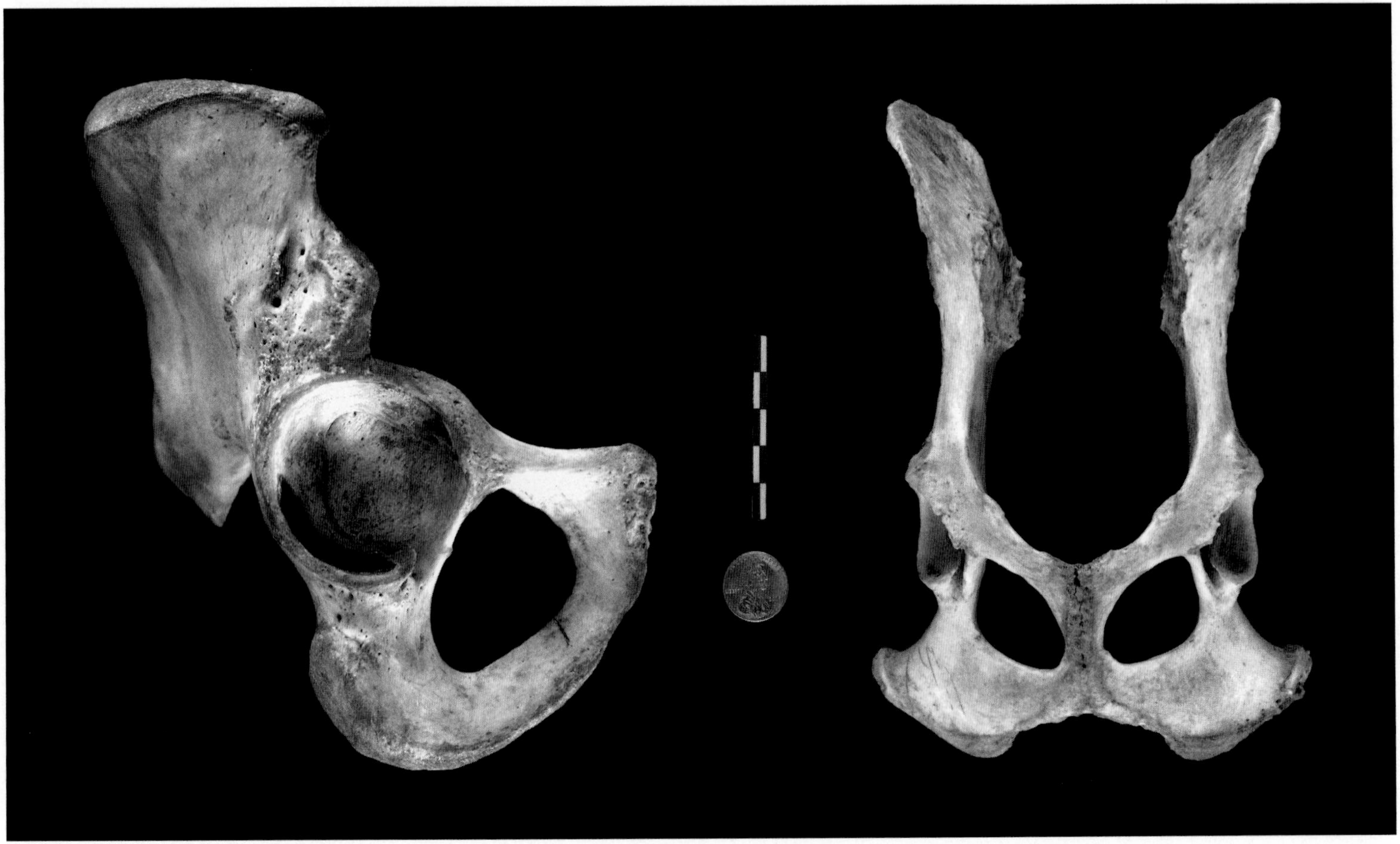

Fig. 9-17. A human right innominate (lateral view) is compared to a dog pelvis (ventral view).

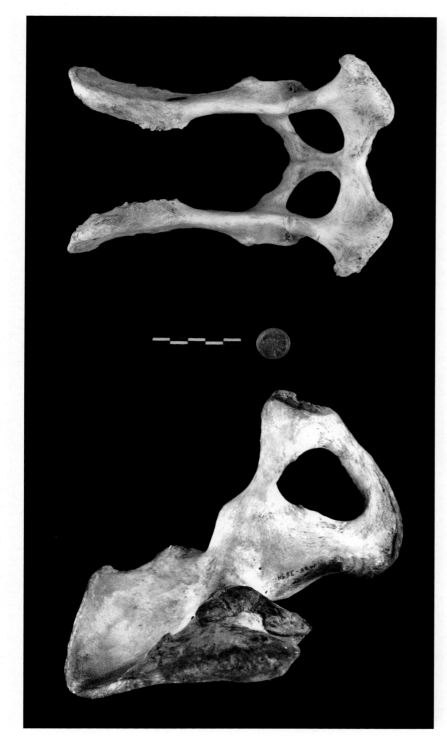

Fig. 9-18. A human left innominate (medial view) is compared to a dog pelvis (dorsal view). The two dog innominates are fused at the pubic symphysis.

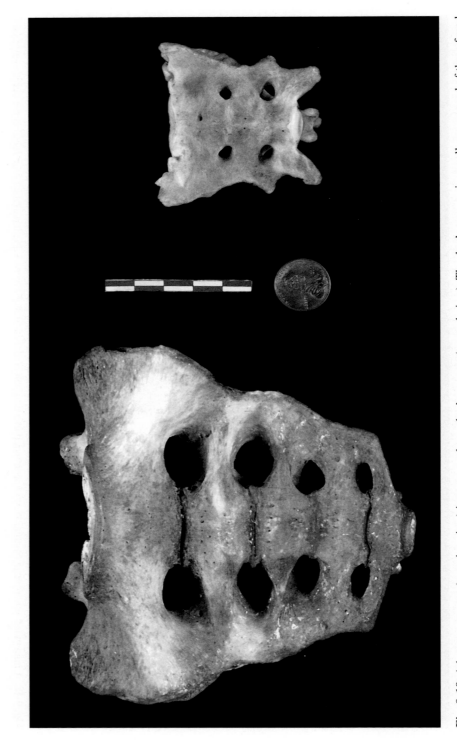

Fig. 9-19. A human sacrum (anterior view) is compared to a dog's sacrum (ventral view). The dog's sacrum is usually composed of three fused vertebrae, while the human's is usually composed of five vertebrae.

172

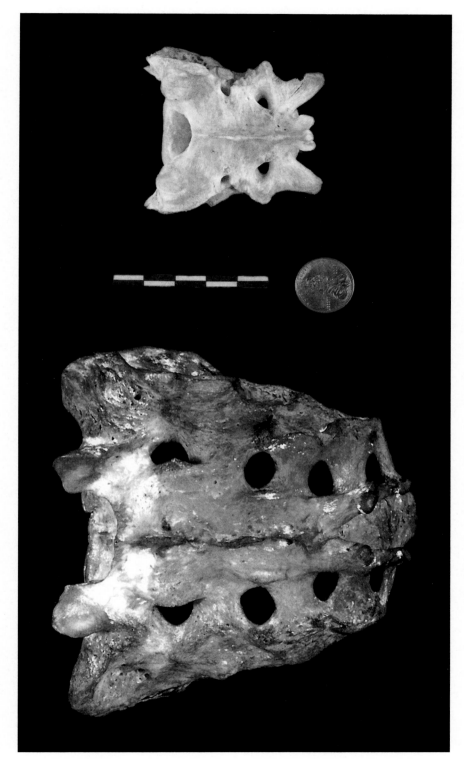

Fig. 9-20. A human sacrum (posterior view) is compared to a dog's sacrum (dorsal view).

173

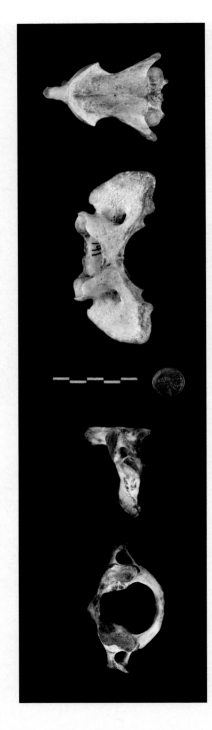

Fig. 9-21. Superior view of human atlas (C1) and lateral view of human axis (C2) compared to those of a dog (ventral views). All mammals have seven cervical vertebrae, and, in general, the length of these vertebrae reflects the length of the animal's neck. The dog's atlas and axis are clearly longer than the human's. The dog atlas is marked by large transverse processes or wings.

174

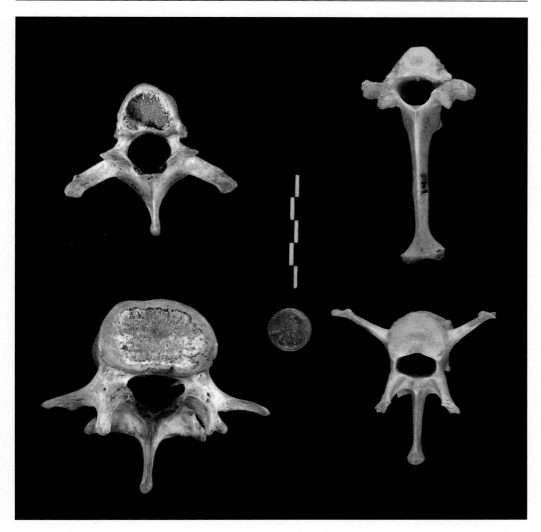

Fig. 9-22. Superior views of typical thoracic (top, left) and lumbar (bottom, left) human vertebrae are compared to cranial views of typical thoracic (top, right) and lumbar (bottom, right) dog vertebrae.

10 Human vs Raccoon

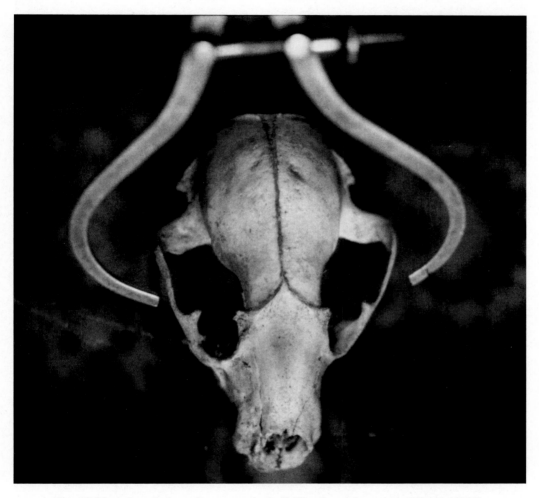

Fig. 10-00. A dorsal view of a raccoon's skull. The dental formula is 3/3.1/1.4/4.2/2.

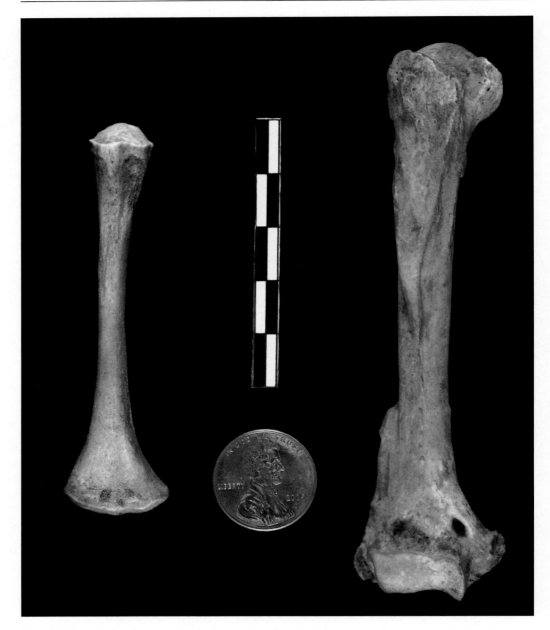

Fig. 10-01. An infant human right humerus (anterior view) is compared to a raccoon's right humerus (cranial view). The raccoon, like the cat, has a supercondylar foramen. The raccoon has a more well-developed lateral epicondylar crest than the cat does.

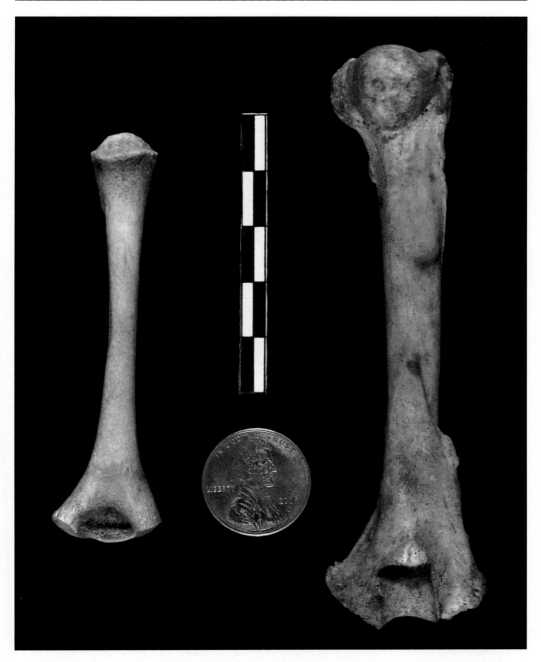

Fig. 10-02. An infant human right humerus (posterior view) is compared to a raccoon's right humerus (caudal view).

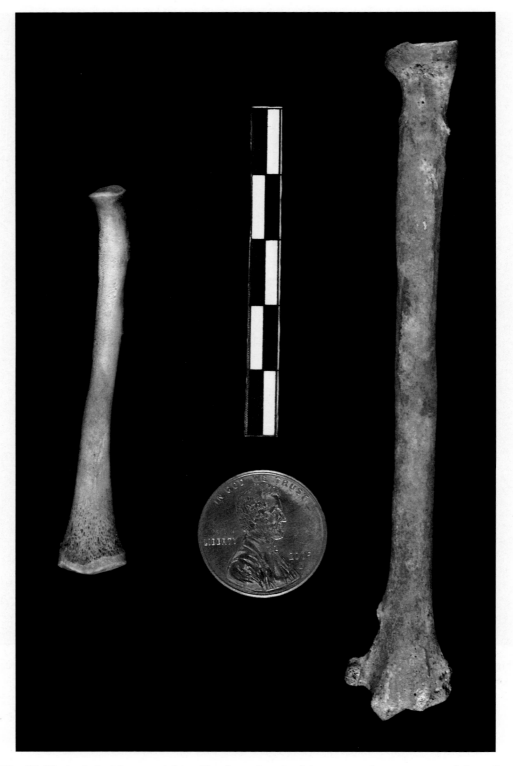

Fig. 10-03. An infant human right radius (anterior view) is compared to a raccoon right radius (caudal view).

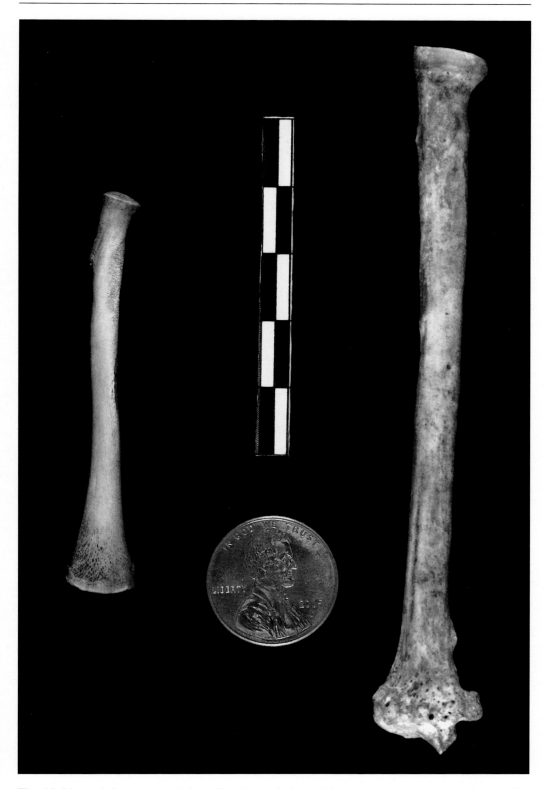

Fig. 10-04. An infant human right radius (posterior view) is compared to a raccoon right radius (cranial view).

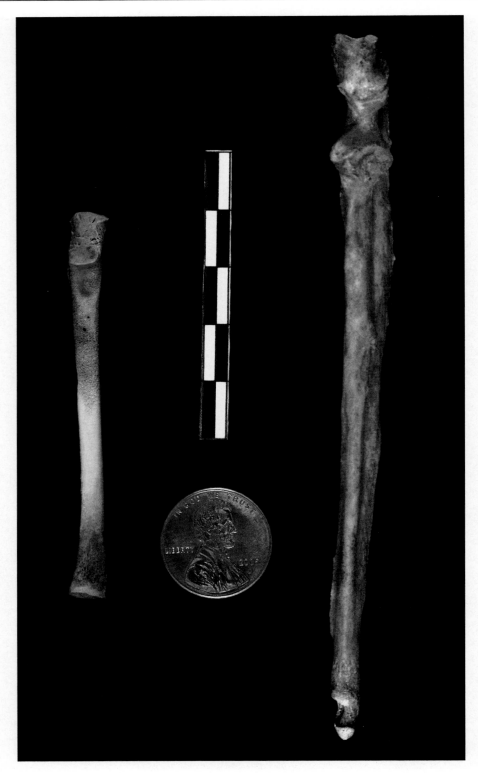

Fig. 10-05. An infant human left ulna (anterior view) is compared to a raccoon's left ulna (cranial view). Note that the distal end of the raccoon's ulna tapers to a small, blunt point, the styloid process.

Fig. 10-06. An infant human left ulna (posterior view) is compared to a raccoon's left ulna (caudal view).

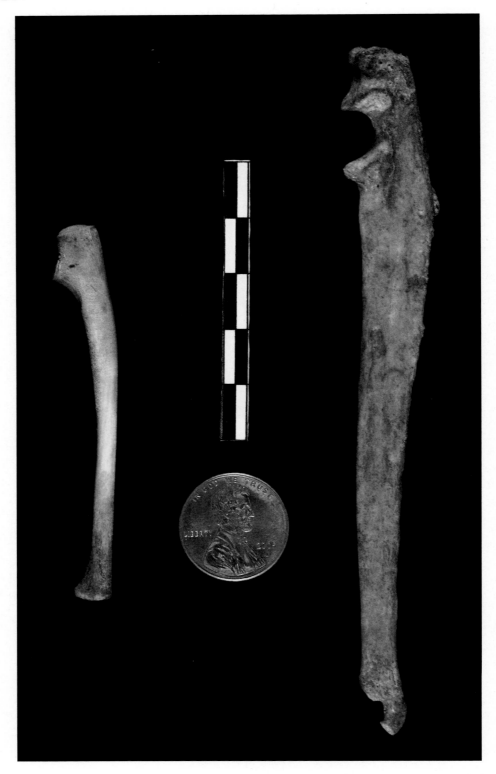

Fig. 10-07. An infant human left ulna (lateral view) is compared to a raccoon's left ulna (lateral view).

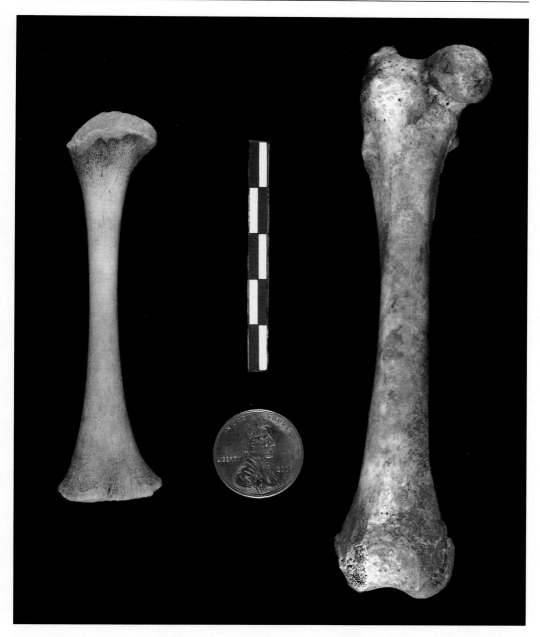

Fig. 10-08. An infant human right femur (anterior view) is compared to a raccoon right femur (cranial view). Note that the second trochanter on the raccoon femur is clearly visible distal to the head.

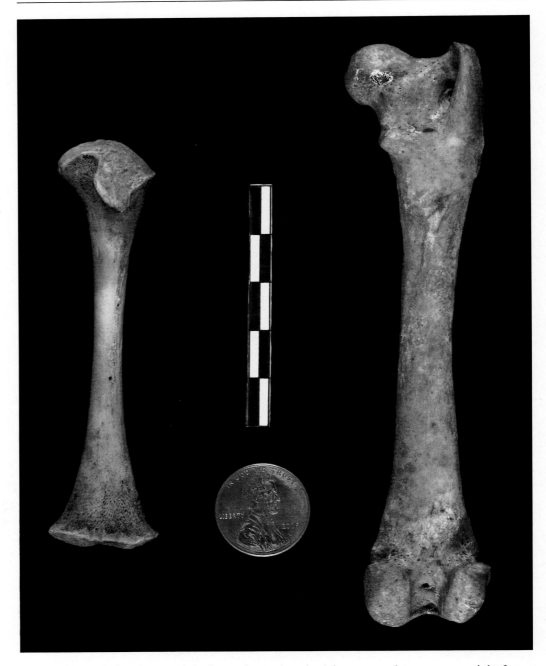

Fig. 10-09. An infant human right femur (posterior view) is compared to a raccoon right femur (caudal view).

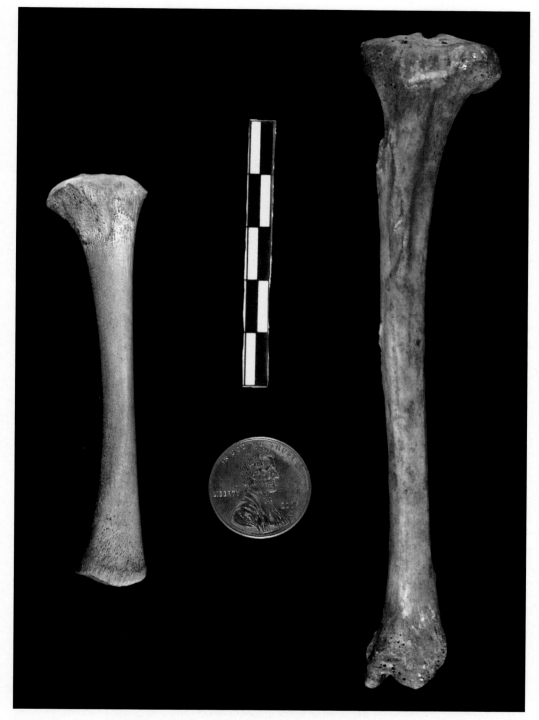

Fig. 10-10. An infant human left tibia (anterior view) is compared to a raccoon's left tibia (cranial view).

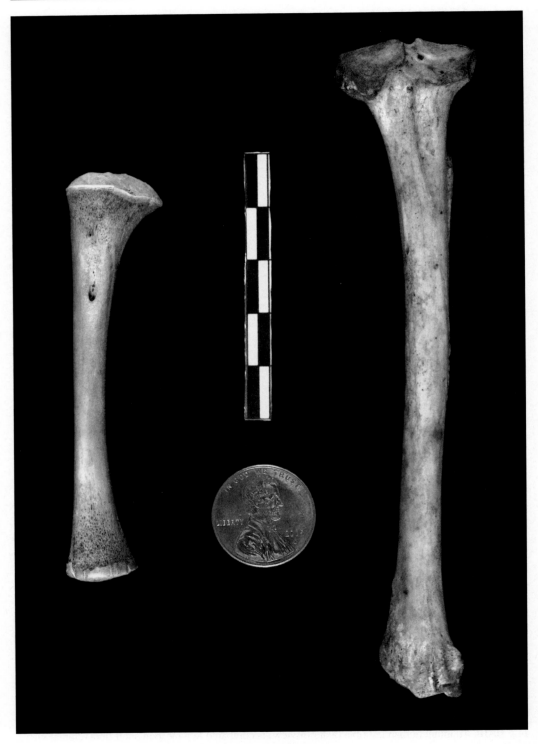

Fig. 10-11. An infant human left tibia (posterior view) is compared to a raccoon's left tibia (caudal view).

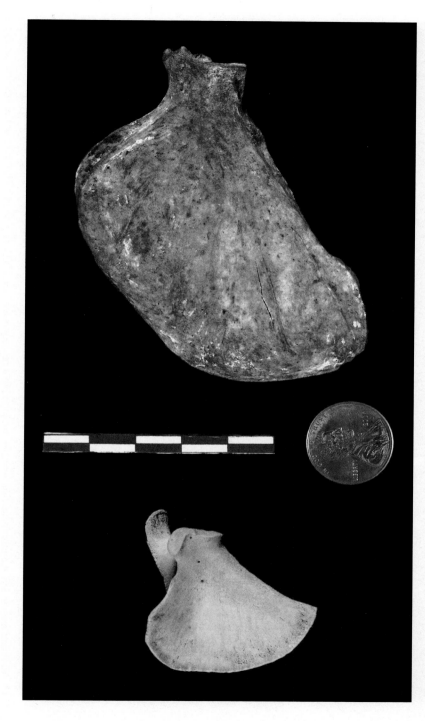

Fig. 10-12. An infant human left scapula (anterior view) is compared to a raccoon's left scapula (medial view). Both are oriented as they would be in a human skeleton. Although the shapes of the two scapulae are similar, the prominent acromion process is visible from the anterior view of the human scapula but not the raccoon.

189

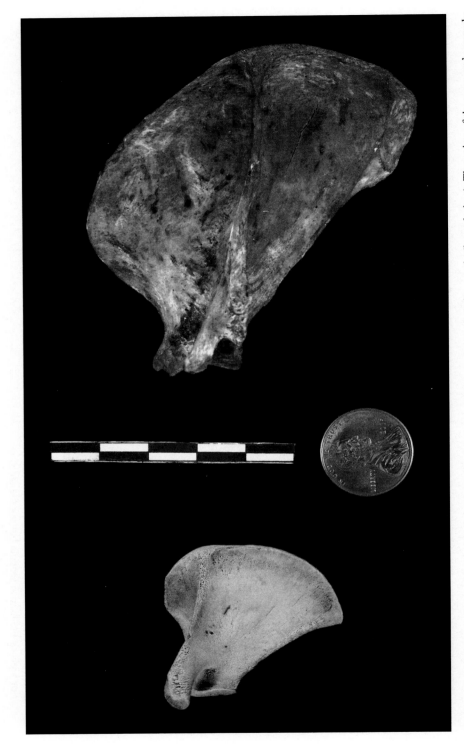

Fig. 10-13. An infant human left scapula (posterior view) is compared to a raccoon's left scapula (lateral view). The spine of the raccoon's scapula divides the scapula neatly in half.

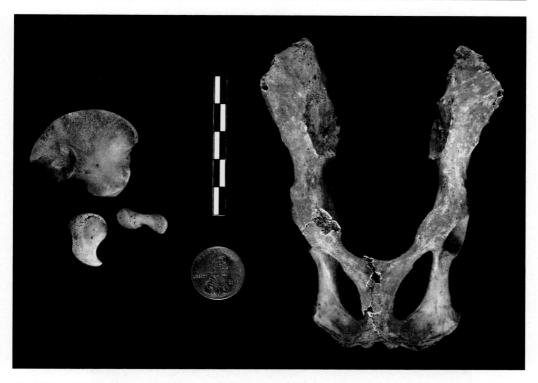

Fig. 10-14. An infant human left innominate (medial view) is compared to a complete raccoon innominate (ventral view). The two halves of the raccoon pelvis have fused along the pubic symphysis. Note the broad blade of the ilium on the human innominate.

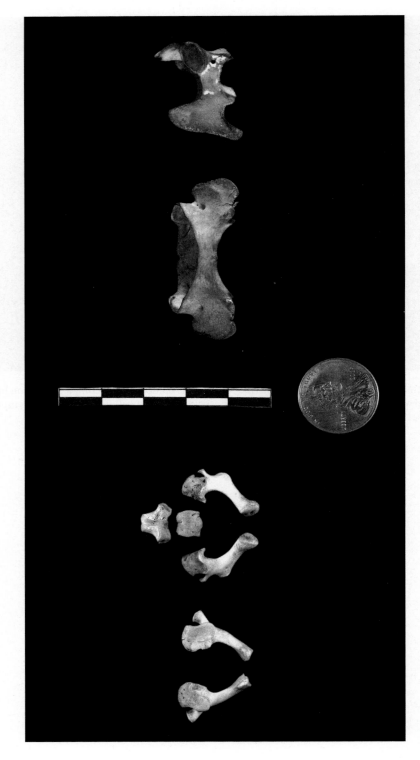

Fig. 10-15. Superior views of an infant human atlas (C1) and axis (C2) are compared to a raccoon atlas (ventral view) and a axis (lateral view).

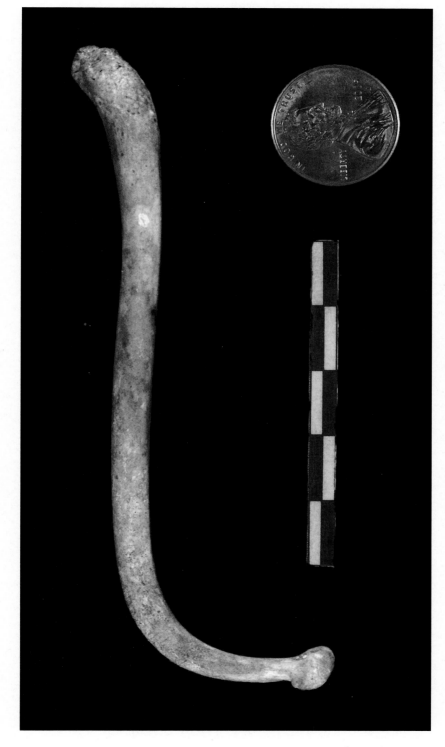

Fig. 10-16. A baculum (penis bone) from a male raccoon.

11 Human vs Opossum

Fig. 11-00. A dorsal view of an opossum skull.

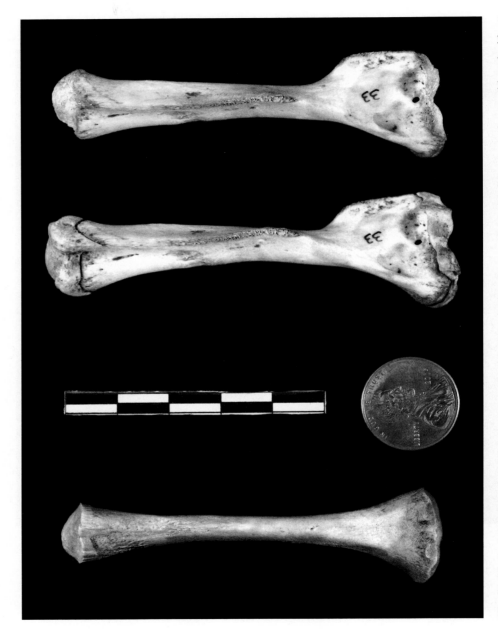

Fig. 11-01. An infant human left humerus (anterior view) is compared to an opossum left humerus shown with and without the epiphyses (cranial views). The possum humerus has a well developed lateral epicondylar crest.

196

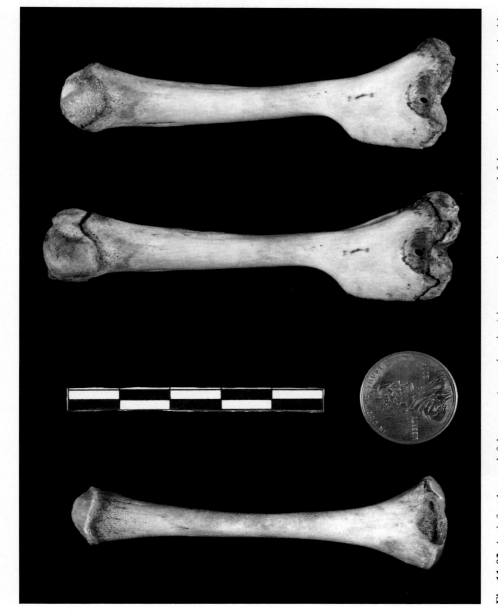

Fig. 11-02. An infant human left humerus (posterior view) is compared to an opossum left humerus shown with and without the epiphyses (caudal views).

197

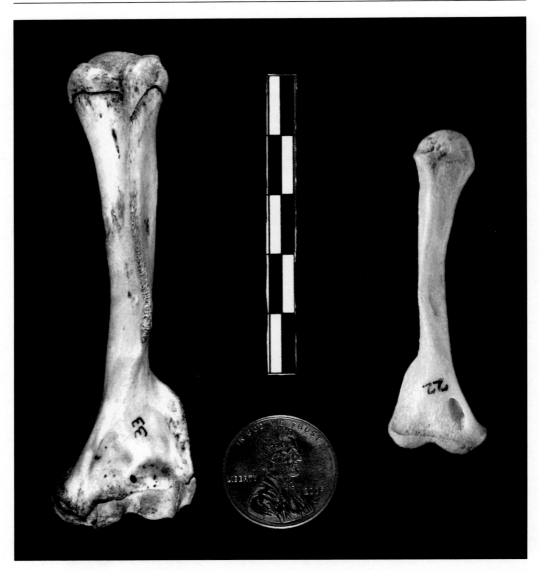

Fig. 11-03. A right opossum humerus (cranial view, with epiphyses) is compared to a left opossum humerus (cranial view, without epiphyses). This photo shows the size and shape differences that may be encountered within the same species.

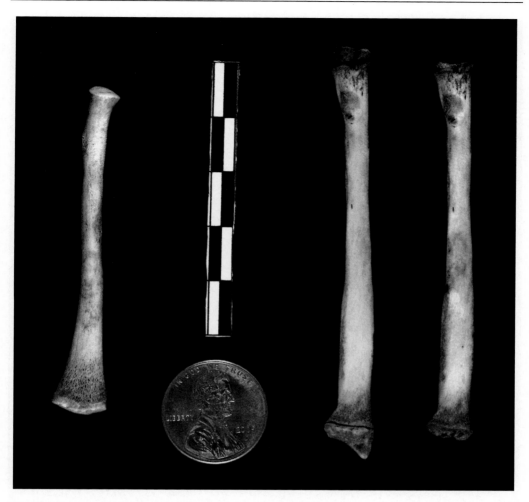

Fig. 11-04. An infant human left radius (anterior view) is compared to an opossum radius shown with and without the epiphyses (caudal view). Human and animal radii are oriented differently. Human skeletons are oriented with the palms up, while animal skeletons are oriented with the paws facing downward.

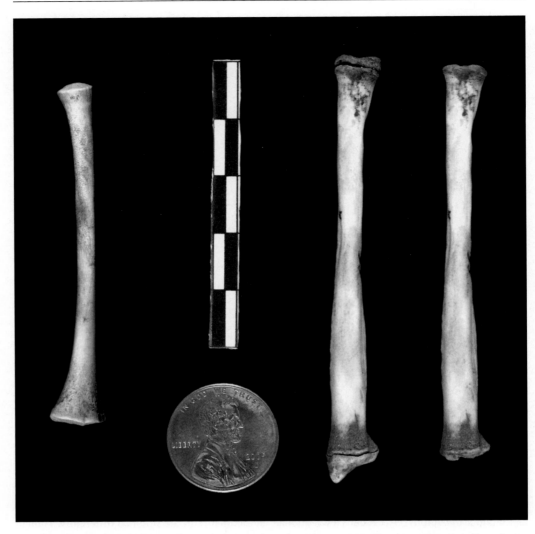

Fig. 11-05. An infant human left radius (posterior view) is compared to an opossum radius shown with and without the epiphyses (cranial view).

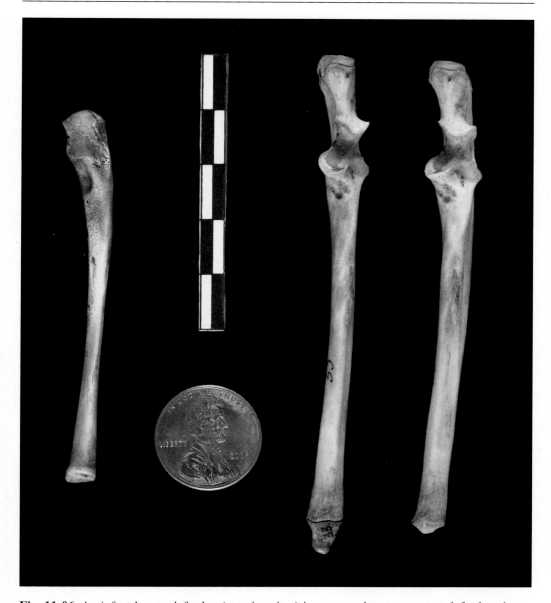

Fig. 11-06. An infant human left ulna (anterior view) is compared to an opossum left ulna shown with and without the epiphyses (cranial views). Note the larger and more well-developed olecranon processes on the opossum ulnae.

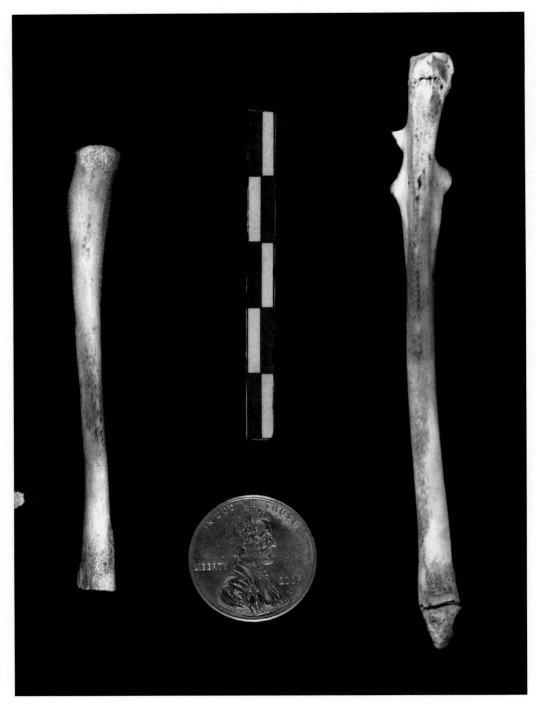

Fig. 11-07. An infant human left ulna (posterior view) is compared to an adult opossum left ulna (caudal view).

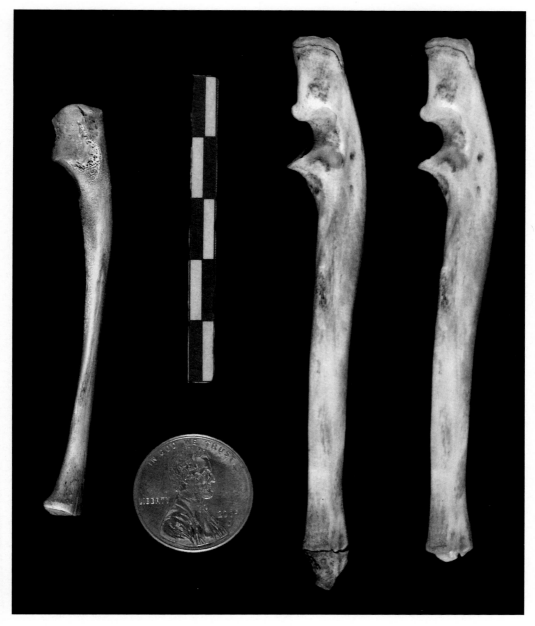

Fig. 11-08. An infant human left ulna (lateral view) is compared to an opossum left ulna shown with and without the epiphyses (lateral views).

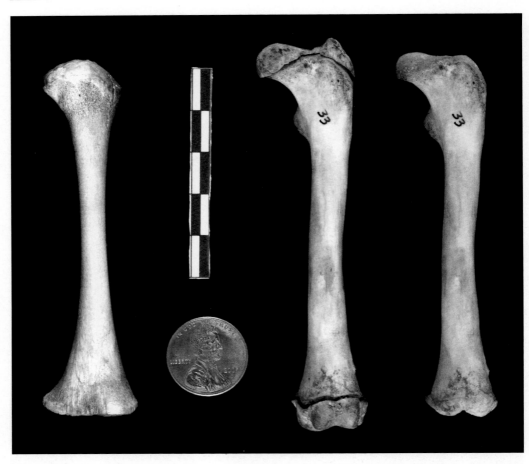

Fig. 11-09. An infant human left femur (anterior view) is compared to an opossum left femur shown with and without the epiphyses (cranial views).

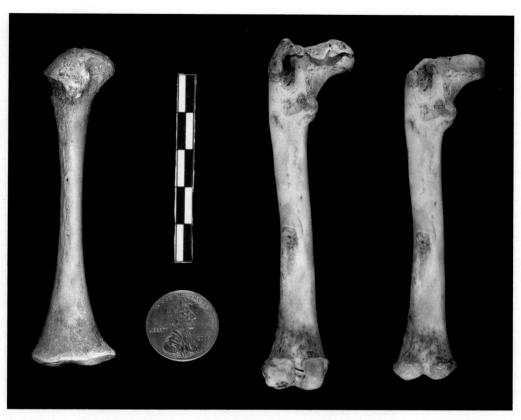

Fig. 11-10. An infant human left femur (posterior view) is compared to an opossum left femur shown with and without the epiphyses (caudal views).

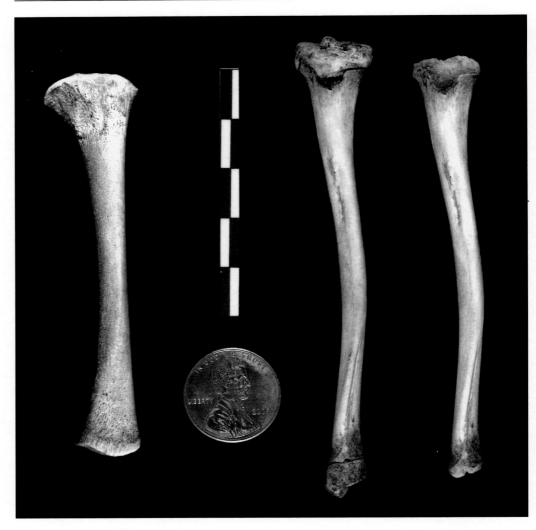

Fig. 11-11. An infant human left tibia (anterior view) is compared to an opossum left tibia shown with and without the epiphyses (cranial views).

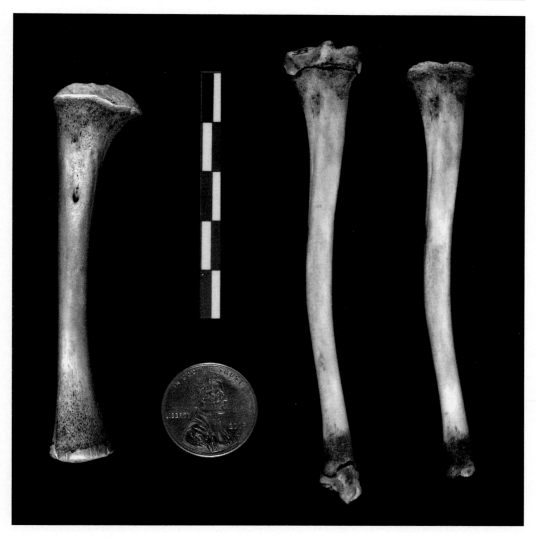

Fig. 11-12. An infant human left tibia (posterior view) is compared to an opossum left tibia shown with and without the epiphyses (caudal views).

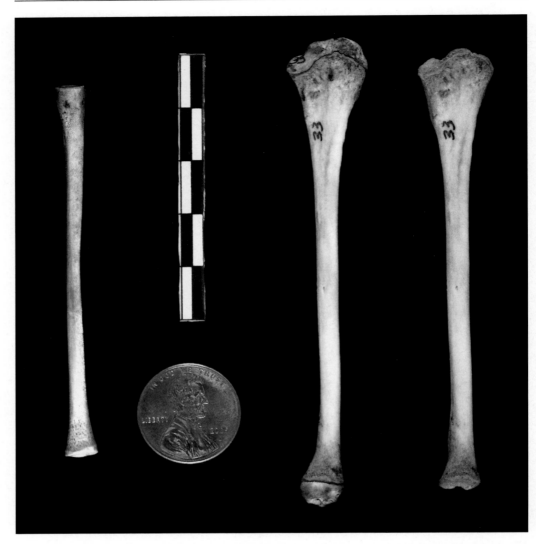

Fig. 11-13. An infant human right fibula (medial view) is compared to an opossum right fibula shown with and without the epiphyses (medial views).

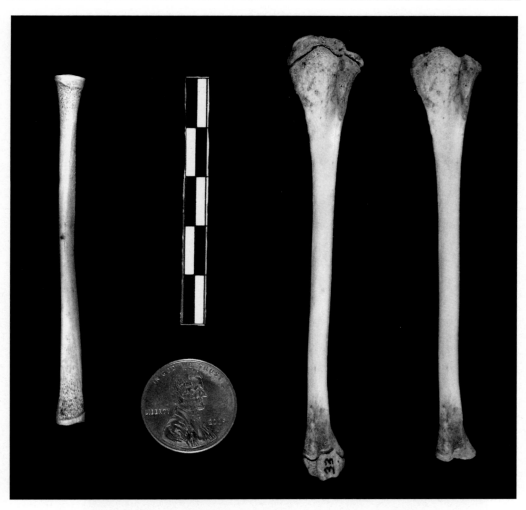

Fig. 11-14. An infant human right fibula (lateral view) is compared to an opossum right fibula shown with and without the epiphyses (lateral views).

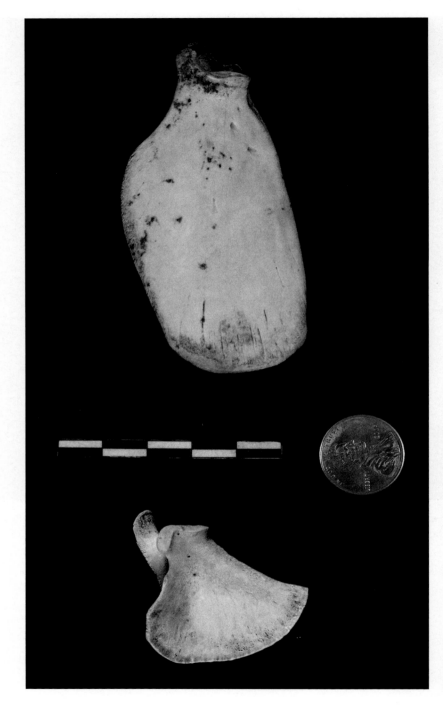

Fig. 11-15. An infant human scapula (anterior view) is compared to a possum scapula (medial view). Both scapulae are oriented as they would be in a human skeleton. Note that the possum scapula is elongated when compared to the human scapula.

210

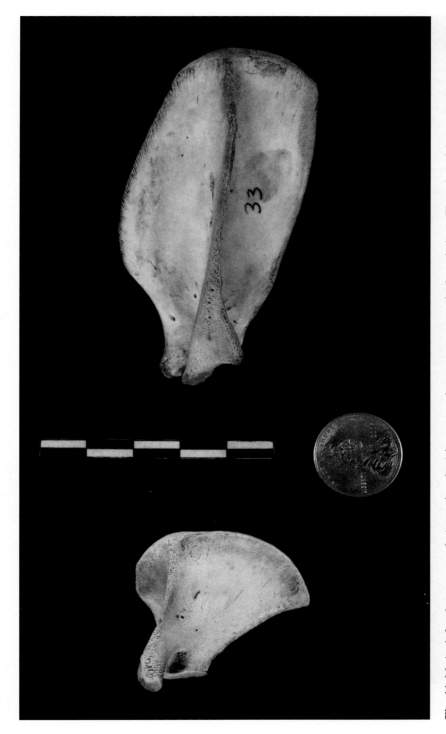

Fig. 11-16. An infant human scapula (posterior view) is compared to a possum scapula (lateral view). The spine of the possum scapula divided the scapula into two equal halves.

211

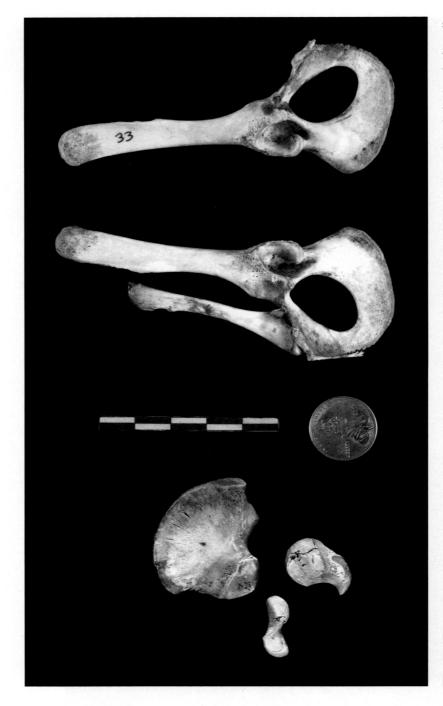

Fig. 11-17. An infant human left innominate (lateral view) is compared to an adult opossum left innominate (lateral view) and a juvenile opossum right opossum (lateral view). The epipubic bone can be seen on the adult innominate.

212

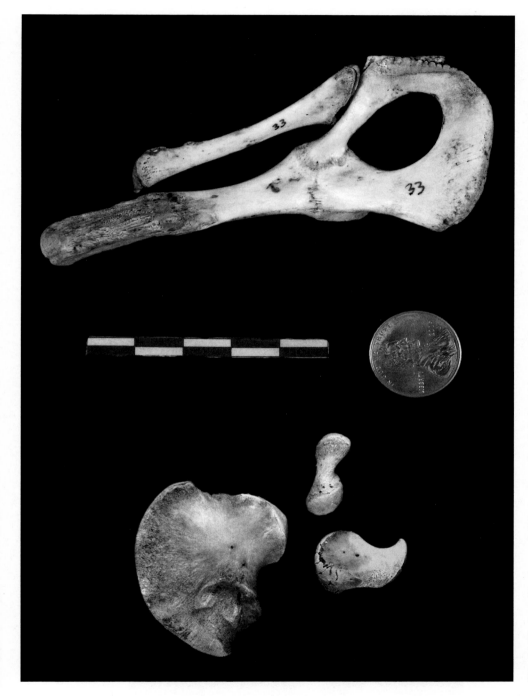

Fig. 11-18. An infant human left innominate (medial view) is compared to an adult opossum left innominate (medial view). The opossum pelvis includes epipubic bones. These bones are found only in marsupials, and they serve to support the pouch.

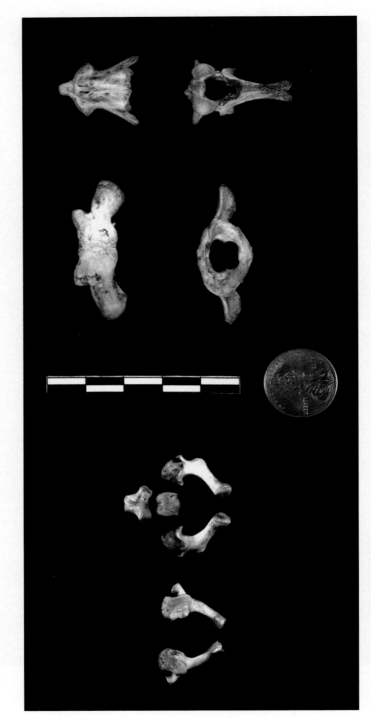

Fig. 11-19. An infant human atlas and axis (C1 and C2, superior views) are compared to an opossum atlas and axis (ventral and cranial views).

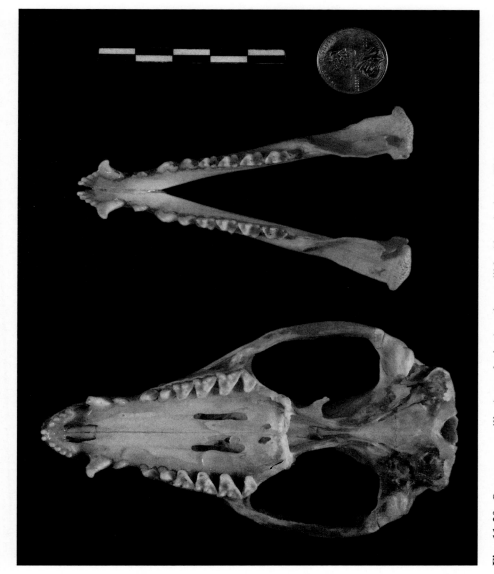

Fig. 11-20. Opossum maxilla (ventral view) and mandible (dorsal view). The opossum dental formula is 5/4.1/1.3/3.4/4. Note that opossums have different numbers of incisors in the upper and lower jaws.

12 Human vs Cat

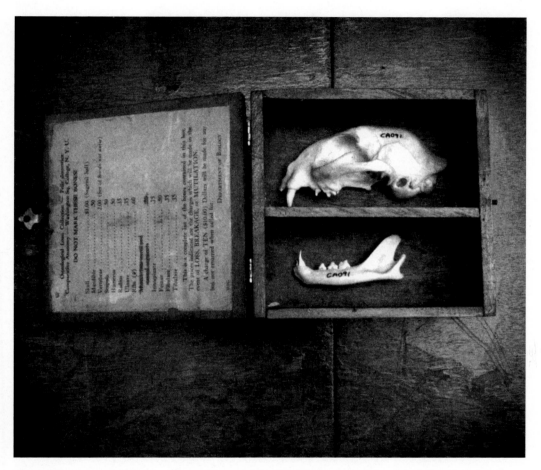

Fig. 12-00. Lateral view of the cat's cranium and mandible. The cat's maxillary dental formula is 3 incisors, 1 canine, 3 premolars, and 1 molar. The mandibular formula is 3 incisors, 1 canine, 2 premolars, and 1 molar.

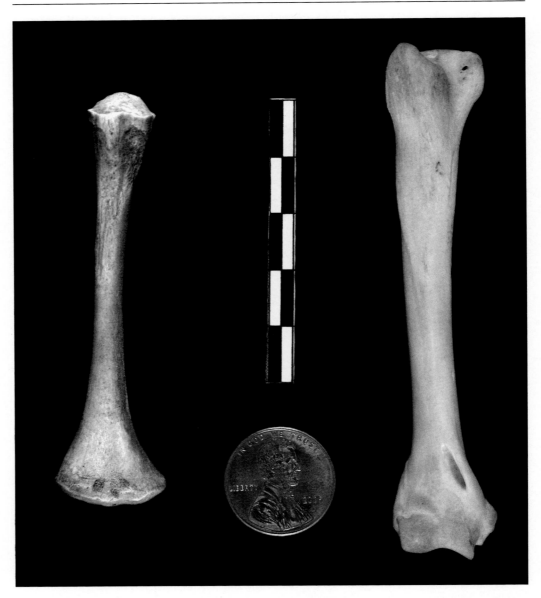

Fig. 12-01. An infant human right humerus (anterior view) is compared to a cat's right humerus (cranial view). The cat humerus includes a supercondylar foramen on the distal portion of the shaft.

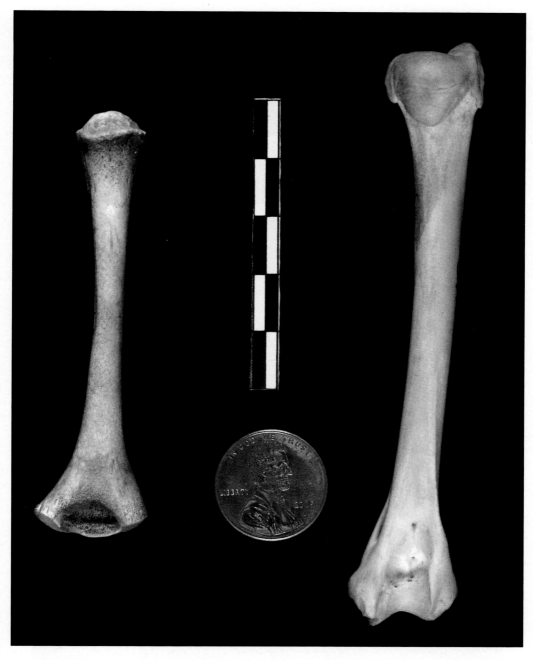

Fig. 12-02. An infant human right humerus (posterior view) is compared to a cat's right humerus (caudal view).

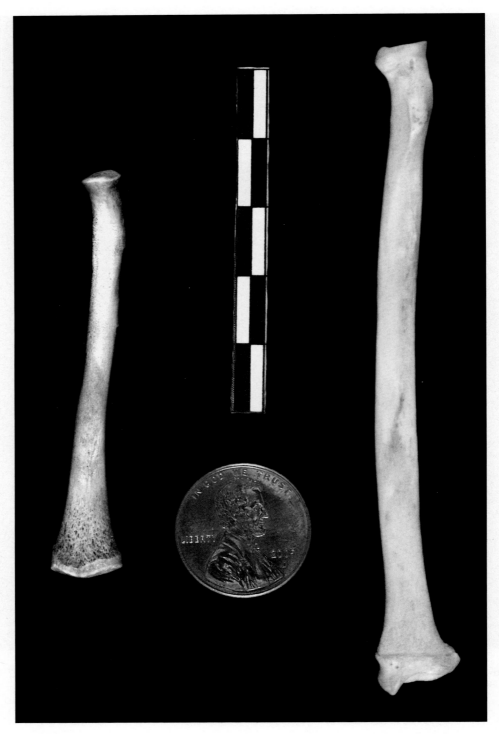

Fig. 12-03. An infant human right radius (anterior view) is compared to a cat's right radius (caudal view). Although the two bones are quite similar, the human radius has a longer neck.

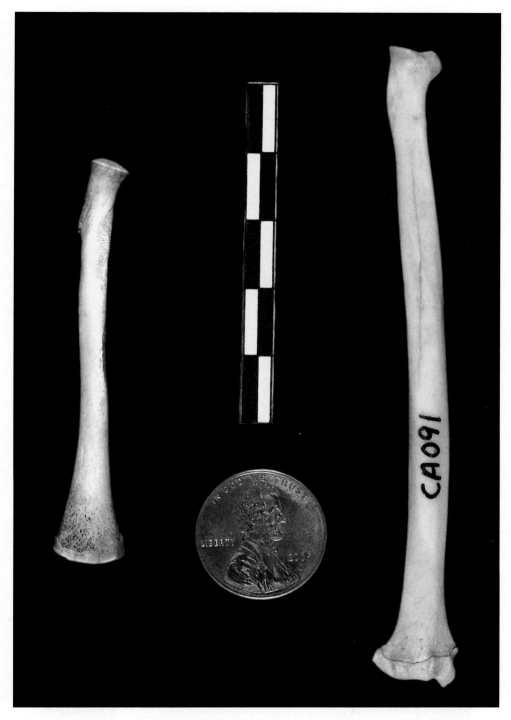

Fig. 12-04. An infant human right radius (posterior view) is compared to a cat's right radius (cranial view).

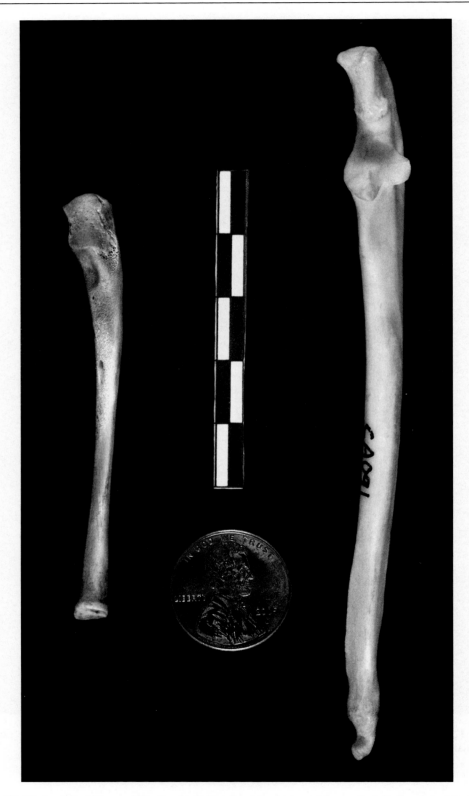

Fig. 12-05. An infant human left ulna (anterior view) is compared to a cat's left ulna (cranial view). Note the larger and more well-developed olecranon process on the cat's ulna.

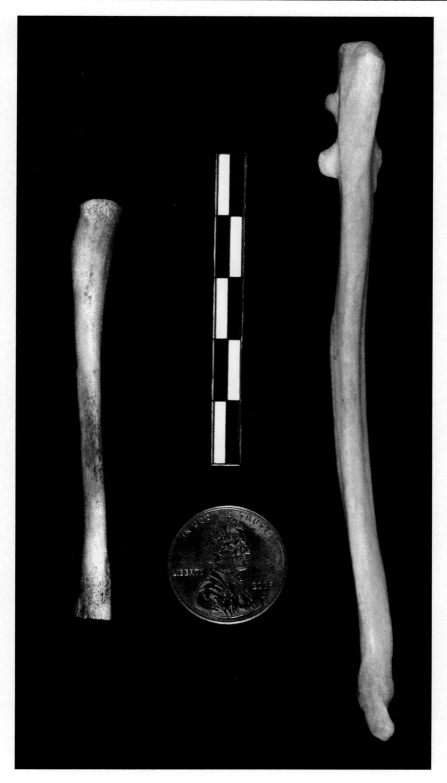

Fig. 12-06. An infant human left ulna (posterior view) is compared to a cat's left ulna (caudal view).

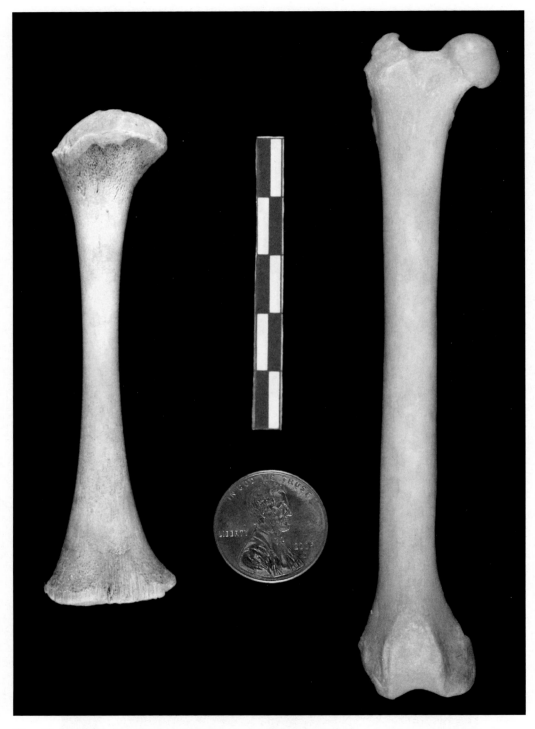

Fig. 12-07. An infant human right femur (anterior view) is compared to a cat's right femur (cranial view).

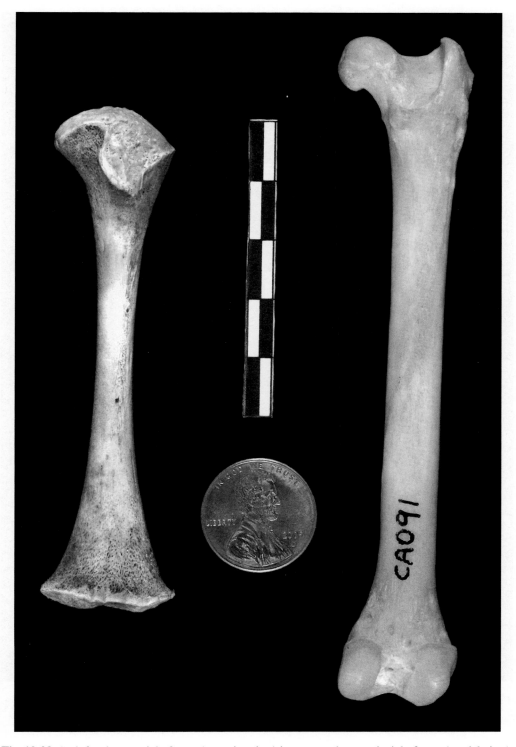

Fig. 12-08. An infant human right femur (posterior view) is compared to a cat's right femur (caudal view).

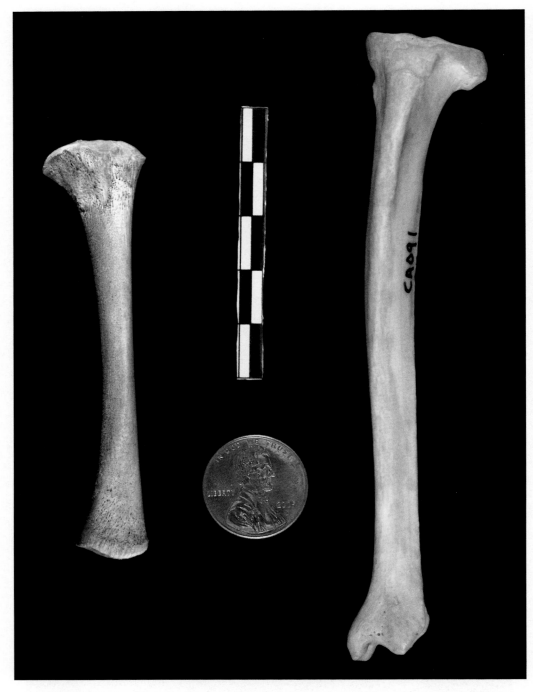

Fig. 12-09. An infant human left tibia (anterior view) is compared to a cat's left tibia (cranial view).

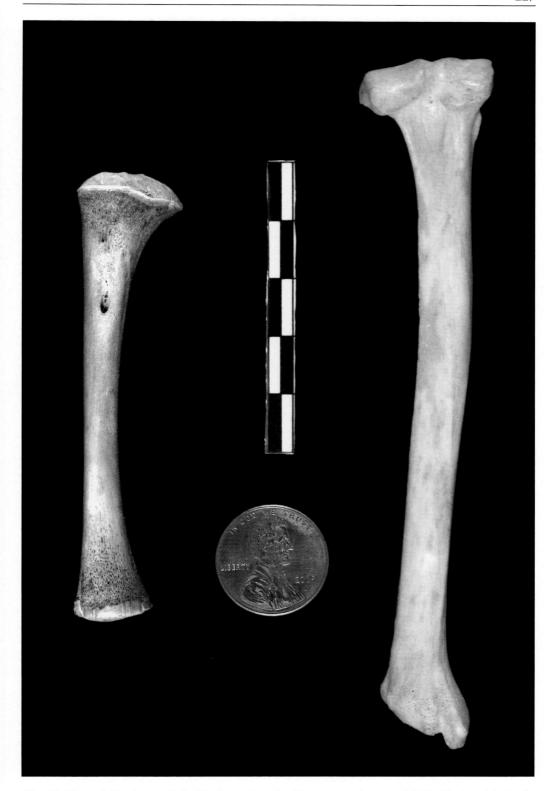

Fig. 12-10. An infant human left tibia (posterior view) is compared to a cat's left tibia (caudal view).

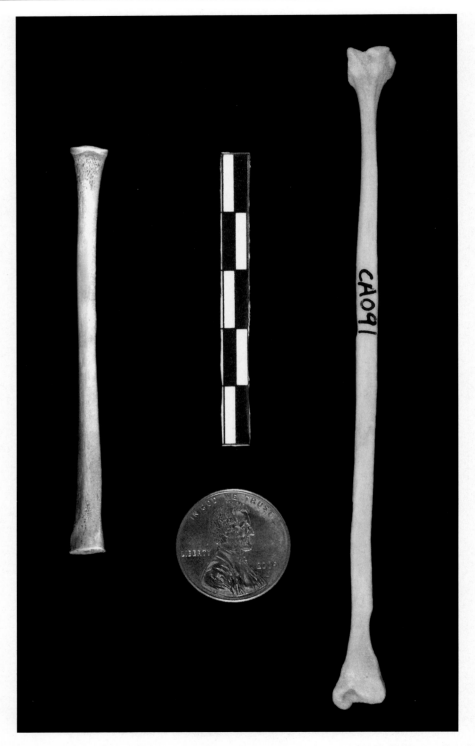

Fig. 12-11. An infant human left fibula (medial view) is compared to a cat's left fibula (medial view).

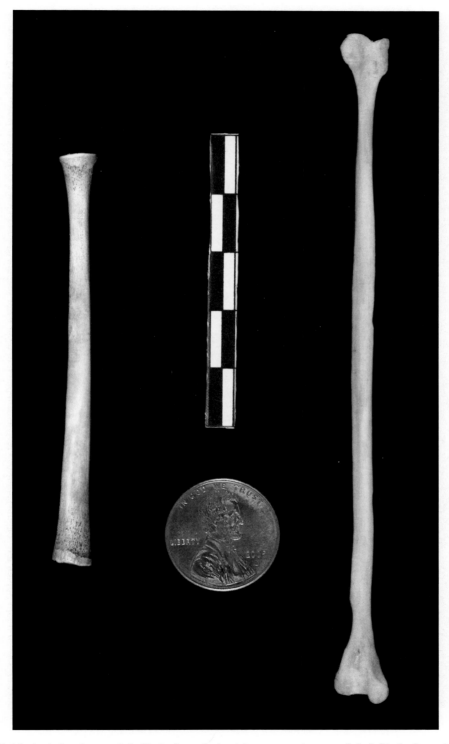

Fig. 12-12. An infant human left fibula (lateral view) is compared to a cat's left fibula (lateral view). The body of the cat's fibula is relatively slender.

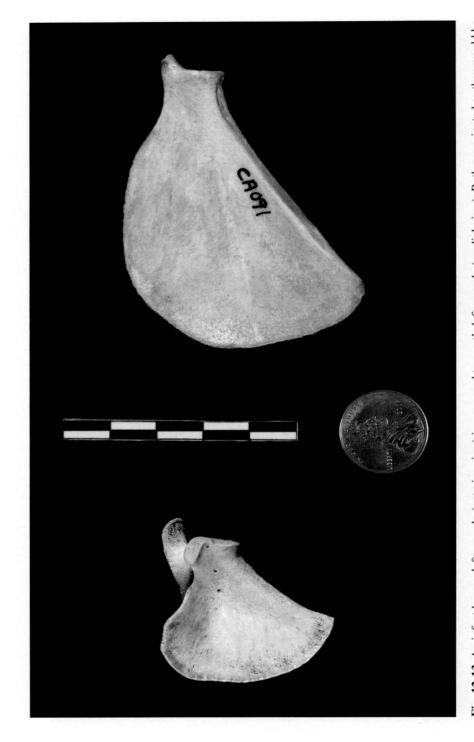

Fig. 12-13. An infant human left scapula (anterior view) is compared to a cat's left scapula (medial view). Both are oriented as they would be in a human. Note the presence of the prominent acromion process on the human scapula which is visible from the anterior view.

Fig. 12-14. An infant human left scapula (posterior view) is compared to a cat's left scapula (lateral view). The cat's scapula is longer that the human scapula.

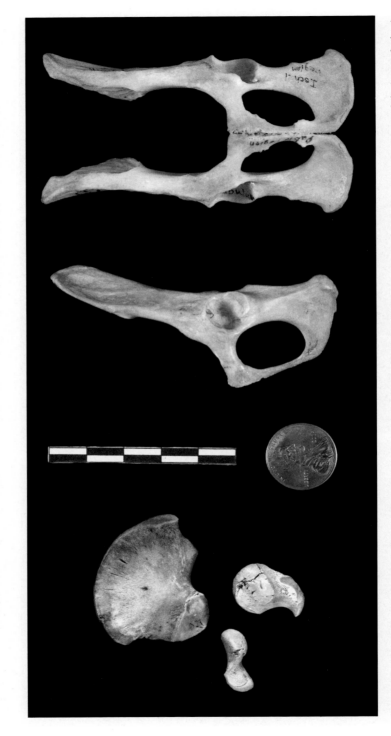

Fig. 12-15. An infant human left pelvis (lateral view) is compared to a cat's left pelvis (lateral view) and a complete cat pelvis (ventral view). Note that the blade of the ilium is much narrower in the cat than in the human.

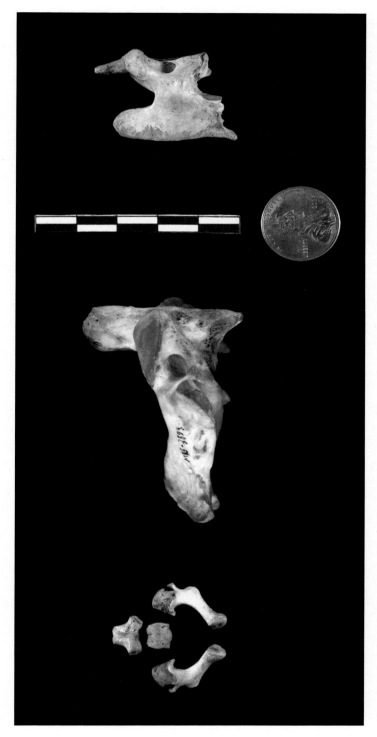

Fig. 12-16. An infant human axis (C2, superior view) and an adult human axis (C2, lateral view) are compared to a cat's axis (lateral view). Note the presence of the large spinous process on the cat's axis.

233

13   Human vs Rabbit

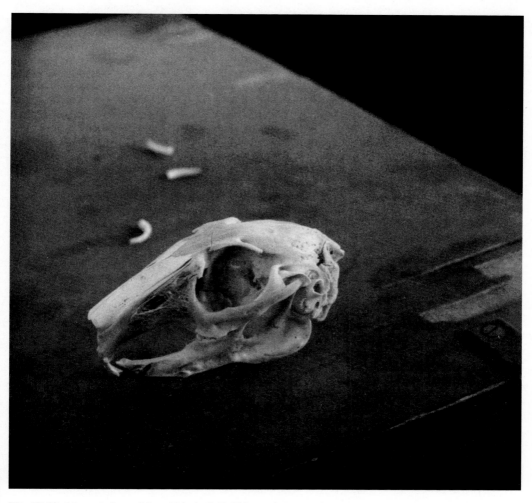

Fig. 13-00. A lateral view of the rabbit skull. Rabbits are lagomorphs and have distinctive dental patterns. The rabbit dental formula is 2/1.0/0.3/2.3/3. The rabbit I^2 is a small peg tooth that is located directly behind the I^1. The rabbit's teeth are open-rooted and continue to grow throughout the animal's lifetime.

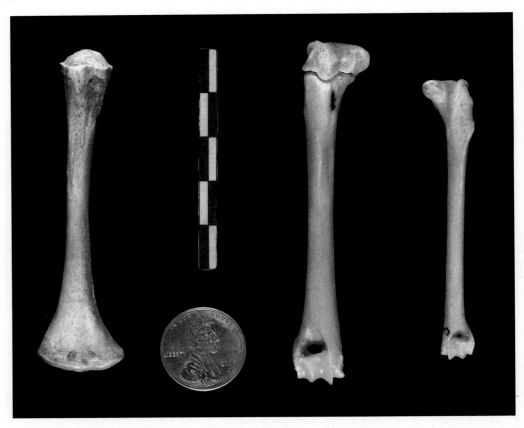

Fig. 13-01. An infant human right humerus (anterior view) is compared to right and left rabbit humeri (cranial views). The rabbit humerus has a supratrochlear foramen.

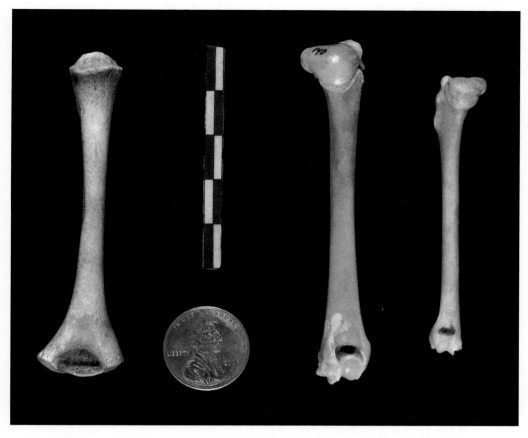

Fig. 13-02. An infant human right humerus (posterior view) is compared to right and left rabbit humeri (caudal views).

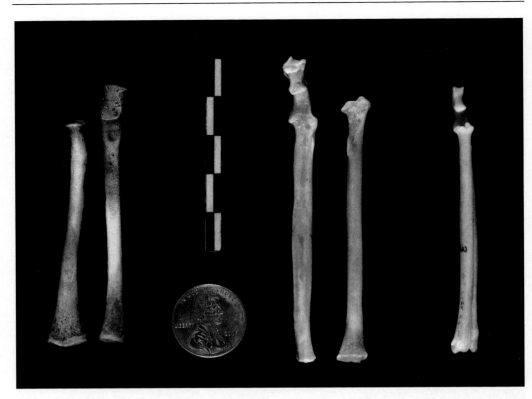

Fig. 13-03. An infant human right radius and ulna (anterior views) are compared to a right rabbit radius (caudal view, separate) and ulna (cranial view, separate) and a left rabbit radius and ulna (cranial view, fused). The human radius includes a distinctive head and neck, while the olecranon process is much larger on the rabbit ulna.

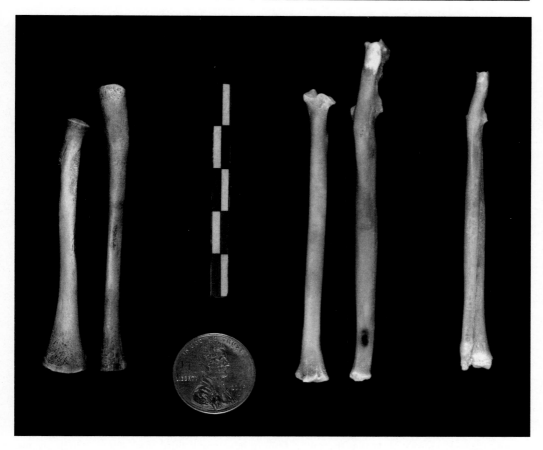

Fig. 13-04. An infant human right radius and ulna (posterior views) are compared to a right rabbit radius (cranial view, separate) and ulna (caudal view, separate) and a left rabbit radius and ulna (caudal view, fused). In the rabbit, the ulna crosses behind the radius and terminates on the lateral side of the carpus.

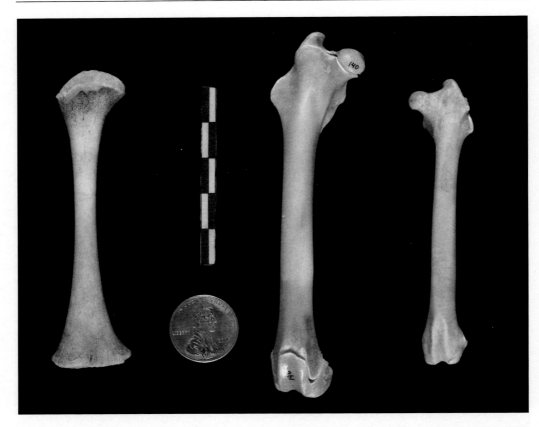

Fig. 13-05. An infant human right femur (anterior view) is compared to right and left rabbit femora (cranial views). Note that the rabbit femora have small third trochanters, which are not seen in the human femur.

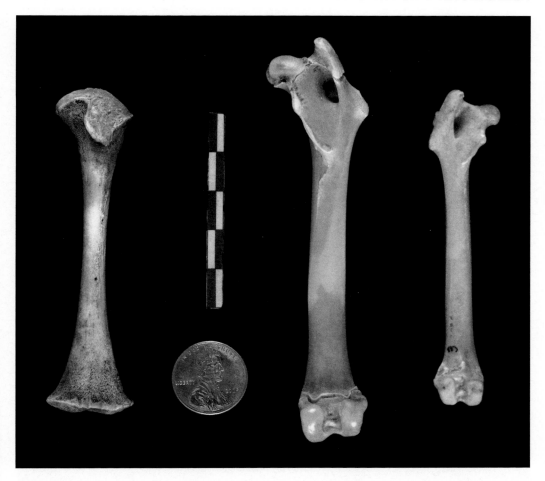

Fig. 13-06. An infant human right femur (posterior view) is compared to right and left rabbit femora (caudal views).

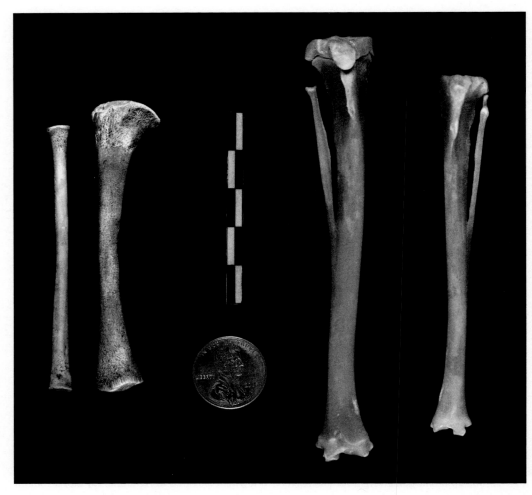

Fig. 13-07. An infant human right tibia and fibula (anterior views) is compared to right and left rabbit tibiae and fibulae (cranial views). The rabbit fibula is fused to the shaft of the tibia.

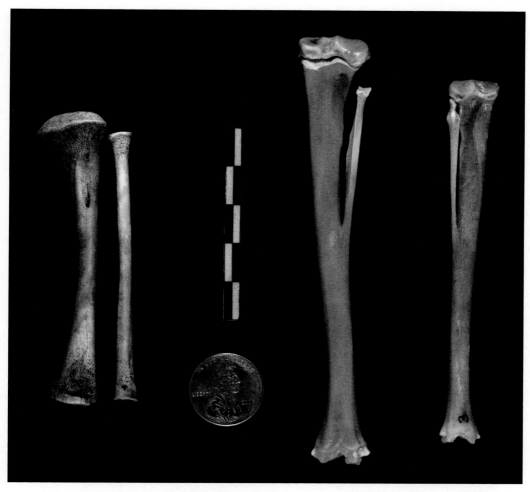

Fig. 13-08. An infant human right tibia and fibula (posterior views) is compared to right and left rabbit tibiae and fibulae (caudal views).

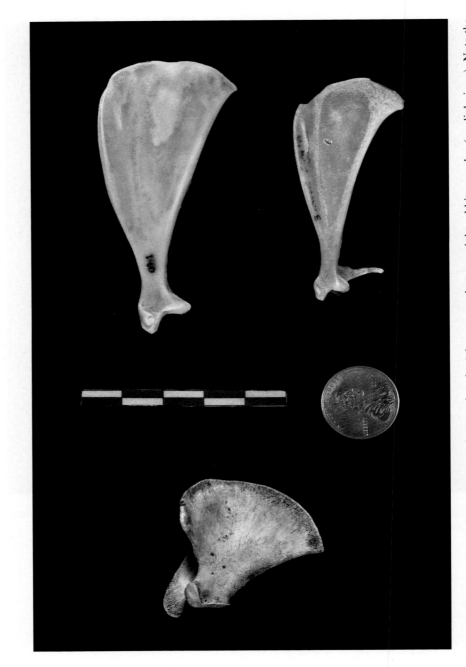

Fig. 13-09. An infant right human scapula (anterior view) is compared to two right rabbit scapulae (medial views). Note the elongated shape of the rabbit scapulae.

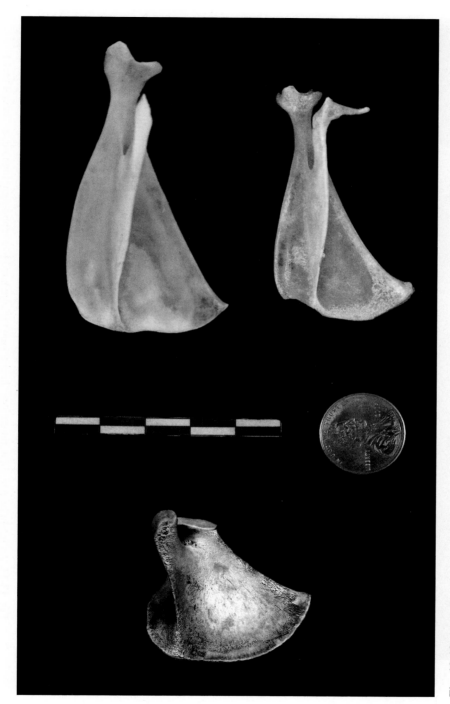

Fig. 13-10. An infant human right scapula (posterior view) is compared to two right rabbit scapulae (lateral views). The rabbit scapula includes a metacromium which is not found on the human scapula. Note that the metacromium is seen more clearly on the smaller rabbit scapula. It is the small process that can be seen near the base of the spine.

245

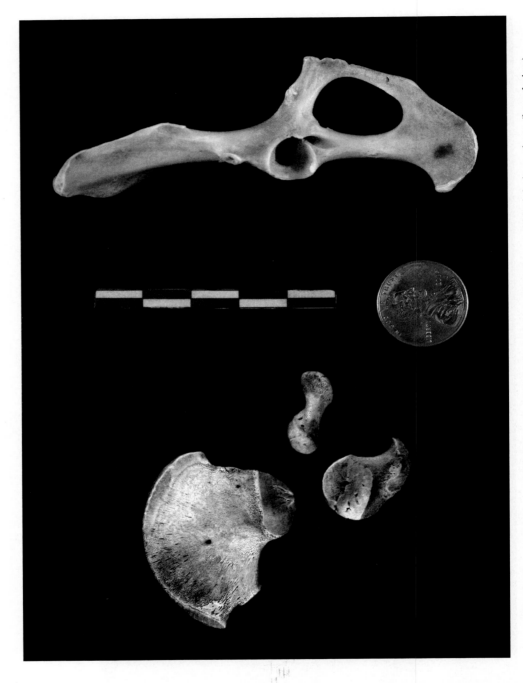

Fig. 13-11. An infant human right innominate (lateral view) is compared to a rabbit right innominate (lateral view).

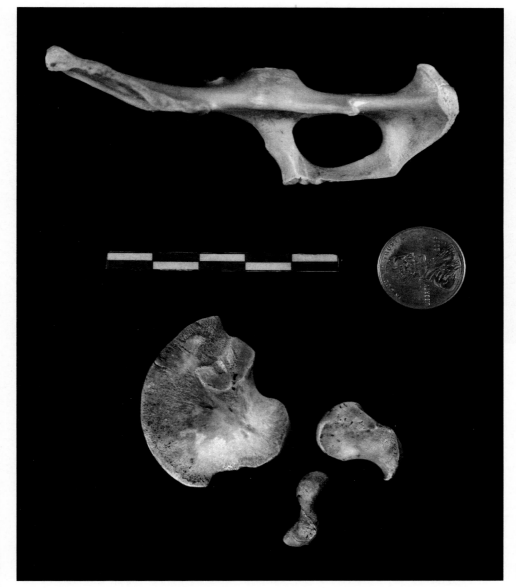

Fig. 13-12. An infant human right innominate (medial view) is compared to a rabbit right innominate (dorsal view).

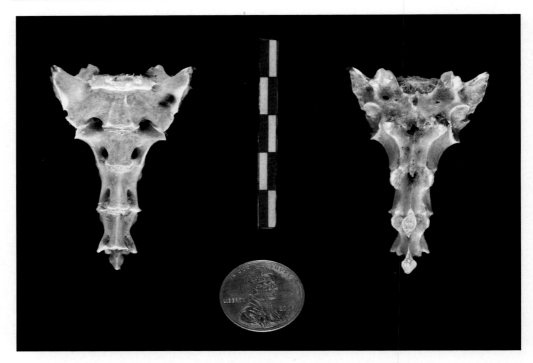

Fig. 13-13. Ventral and dorsal views of the rabbit sacrum.

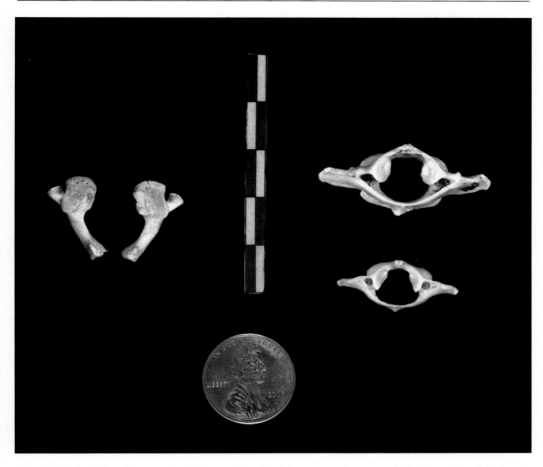

Fig. 13-14. An infant human atlas (C1, superior view) is compared to two rabbit atlases (cranial views).

14 Human vs Turkey

Fig. 14-00. Turkey skull with associated vertebrae.

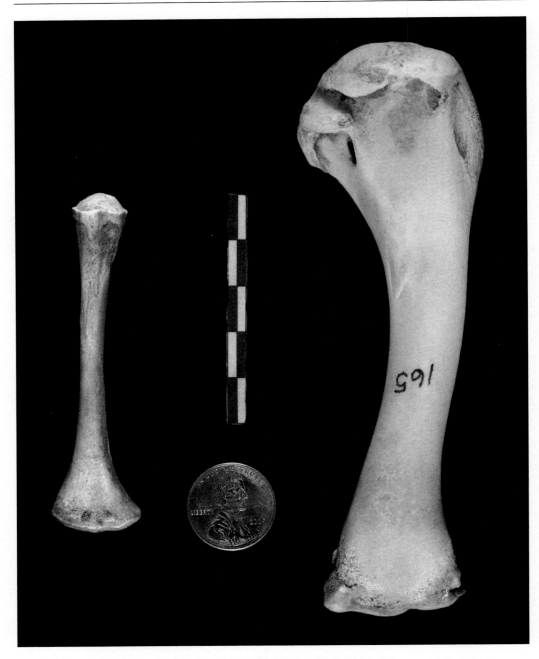

Fig. 14-01. An infant human right humerus (anterior view) is compared to a turkey's right humerus (cranial view). Note the pneumatic fossa near the proximal end of the turkey's humerus. The pneumatic fossa allows for the invasion of the clavicular air sac which pneumatizes the interior of the humerus. This lightens the skeleton, which is an adaptation for flight.

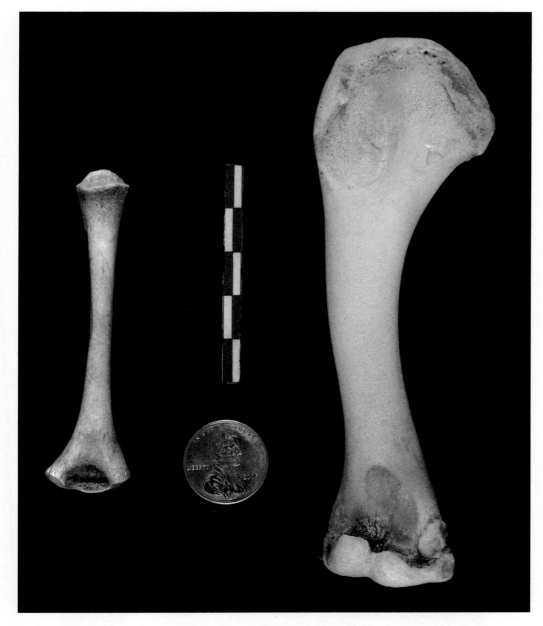

Fig. 14-02. An infant human right humerus (posterior view) is compared to a turkey's left humerus (caudal view).

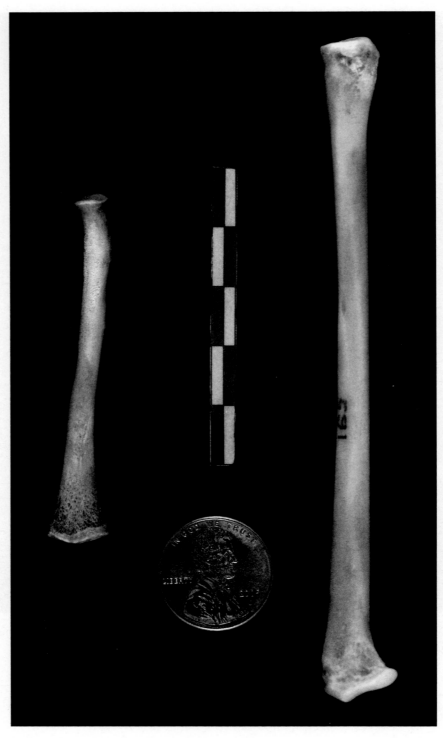

Fig. 14-03. An infant human right radius (anterior view) is compared to a turkey's right radius (cranial view).

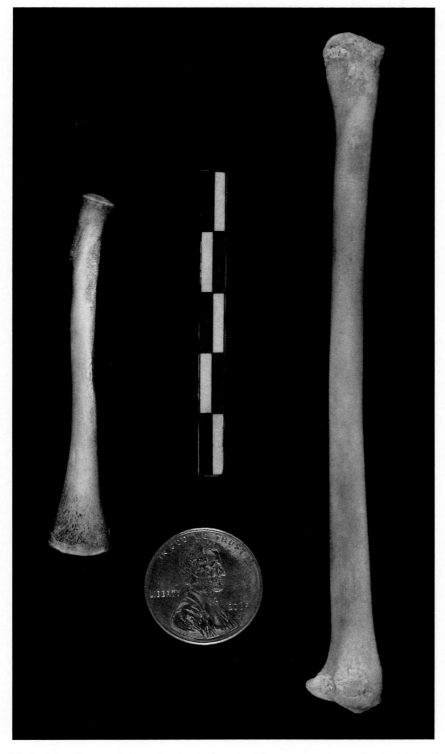

Fig. 14-04. An infant human right radius (posterior view) is compared to a turkey's right radius (caudal view).

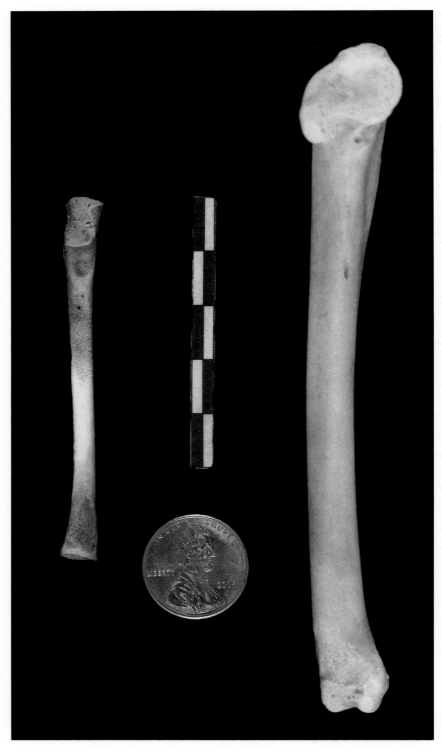

Fig. 14-05. An infant human right ulna (anterior view) is compared to a turkey's right ulna (cranial view).

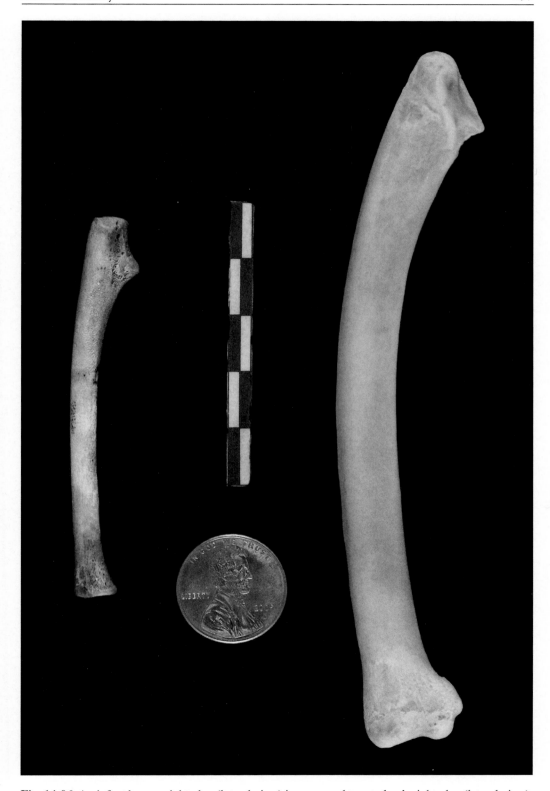

Fig. 14-06. An infant human right ulna (lateral view) is compared to a turkey's right ulna (lateral view).

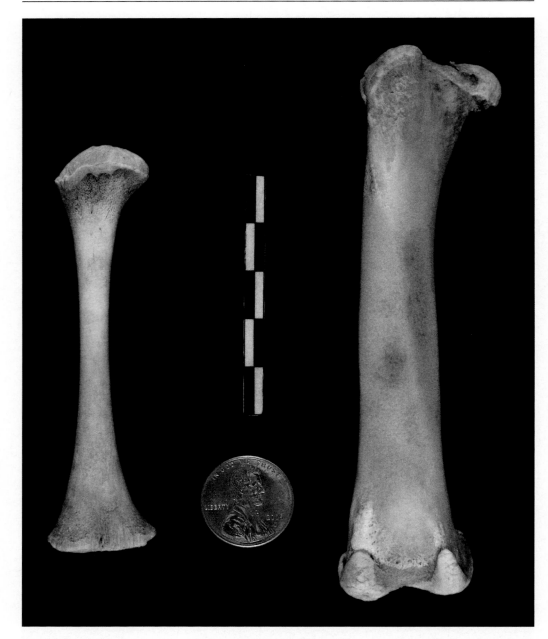

Fig. 14-07. An infant human right femur (anterior view) is compared to a turkey's right femur (cranial view).

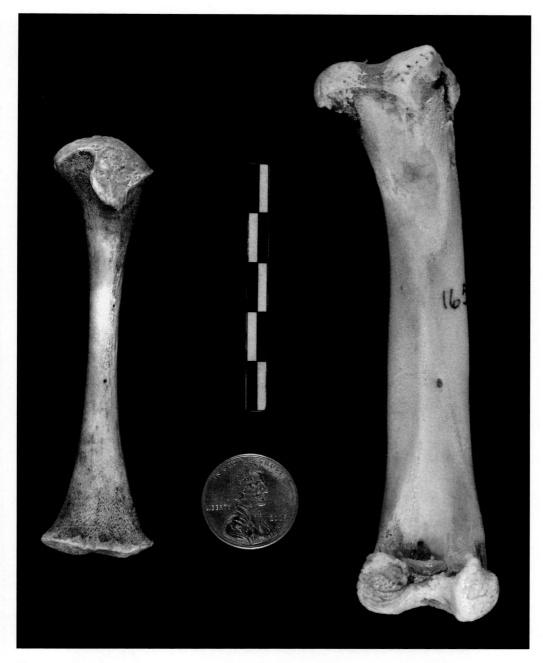

Fig. 14-08. An infant human right femur (posterior view) is compared to a turkey's right femur (caudal view).

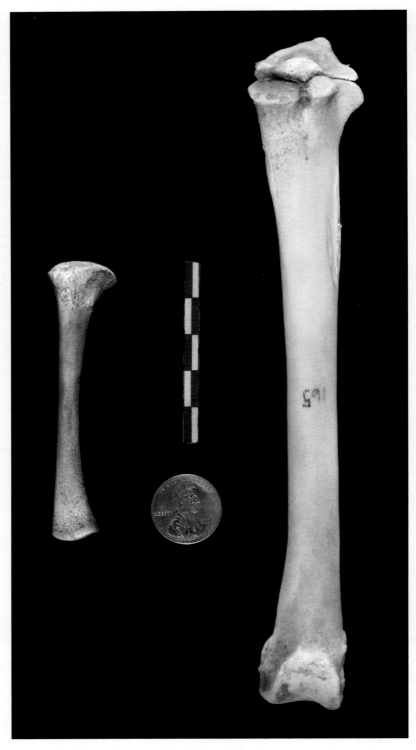

Fig. 14-09. An infant human right tibia (anterior view) is compared to a turkey's right tibiotarsus (cranial view).

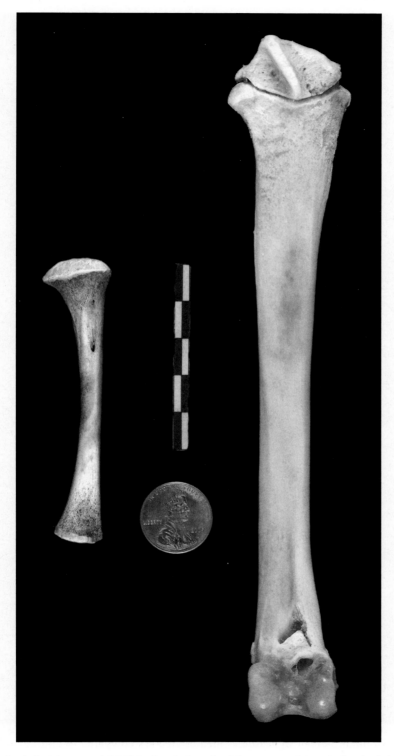

Fig. 14-10. An infant human right tibia (posterior view) is compared to a turkey's right tibiotarsus (caudal view).

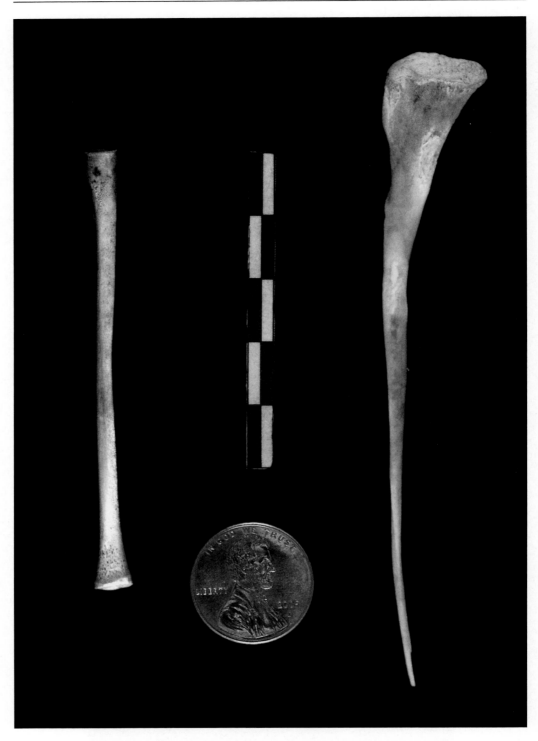

Fig. 14-11. An infant human right fibula (medial view) is compared to a turkey's right fibula (medial view).

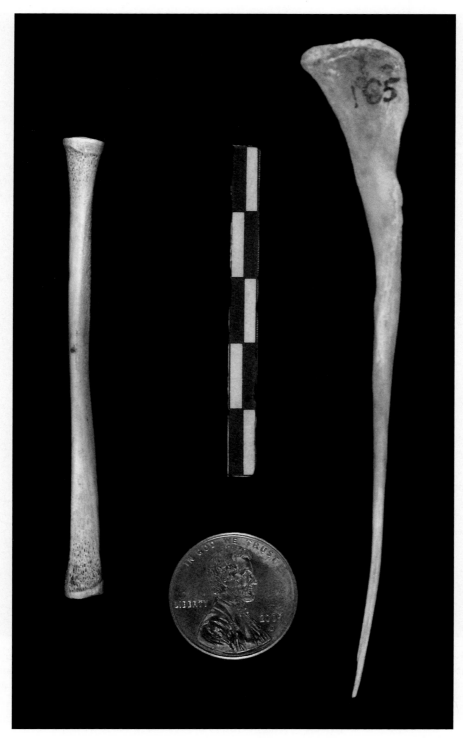

Fig. 14-12. An infant human right fibula (lateral view) is compared to a turkey's right fibula (lateral view).

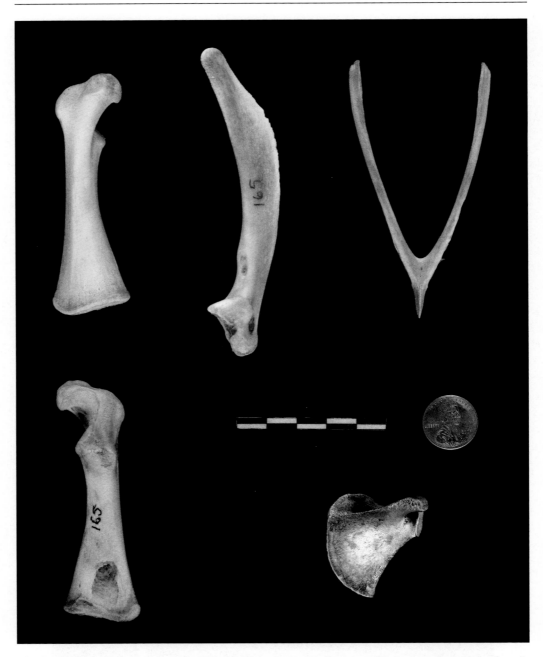

Fig. 14-13. The bird pectoral girdle includes three elements: the coracoid, the scapula, and the furcula (wishbone). A turkey's right coracoid (ventral and dorsal views) is pictured on the left top and bottom, a turkey's scapula (costal view) is pictured on the top middle, and a turkey's furcula is pictured on the top right. The bottom right of the photo shows an infant human right scapula (posterior view).

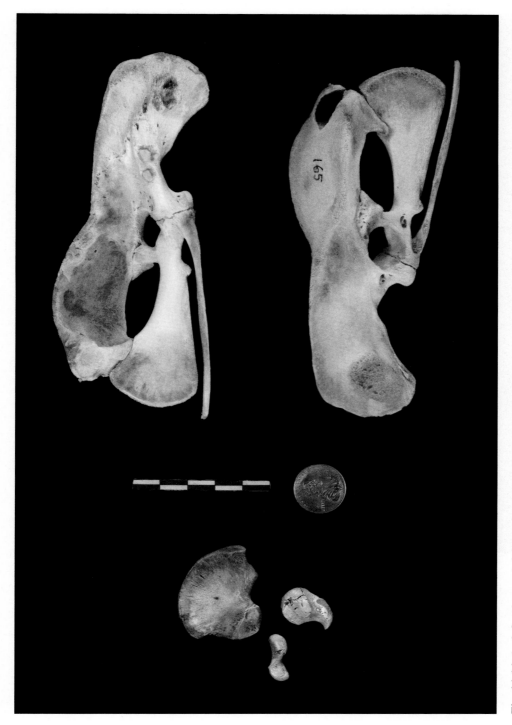

Fig. 14-14. An infant human left innominate (lateral view) is compared to a turkey's left pelvis (medial and lateral views). Note that the turkey's acetabulum is perforated.

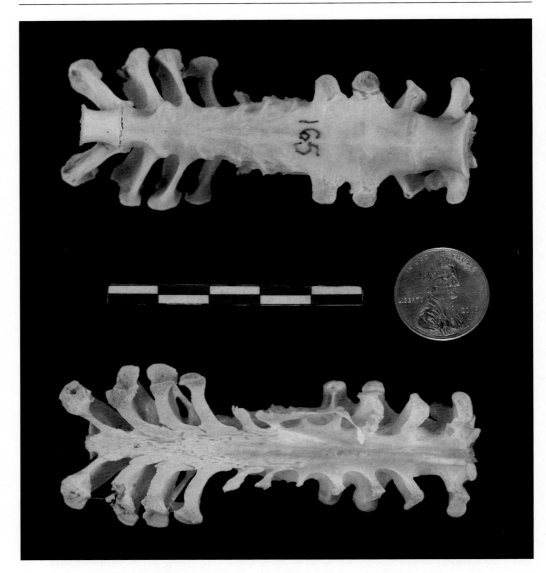

Fig. 14-15. Ventral and dorsal views of a turkey's synsacrum.

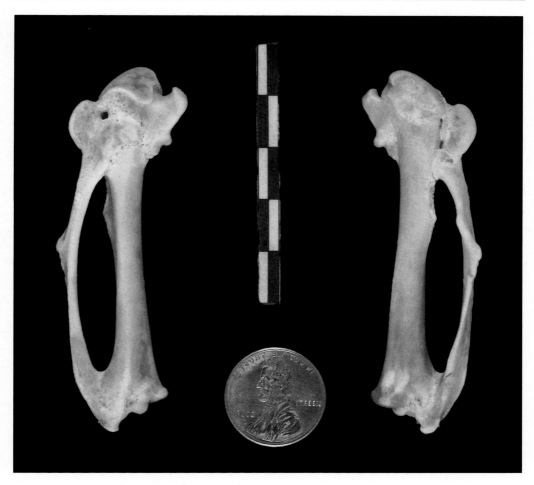

Fig. 14-16. Turkey right carpometacarpus (dorsal and ventral views). The bone is made up of three fused elements-the 2nd, 3rd, and 4th metacarpals.

15 Human vs Duck

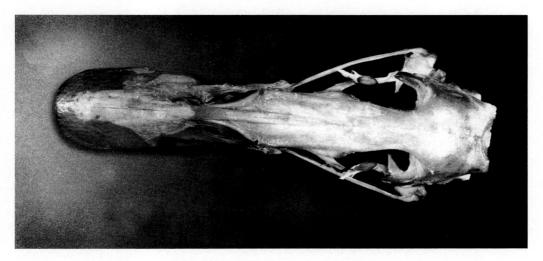

Fig. 15-00. A dorsal view of a duck's skull.

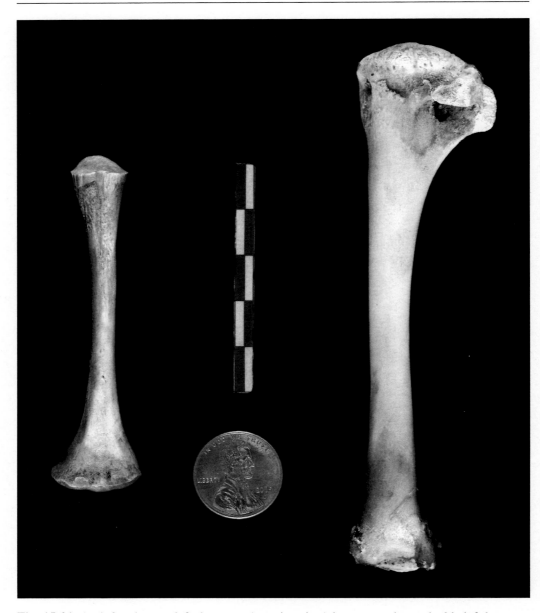

Fig. 15-01. An infant human left humerus (anterior view) is compared to a duck's left humerus (cranial view). Note the pneumatic fossa near the proximal end of the duck's humerus. This is an adaptation for flight.

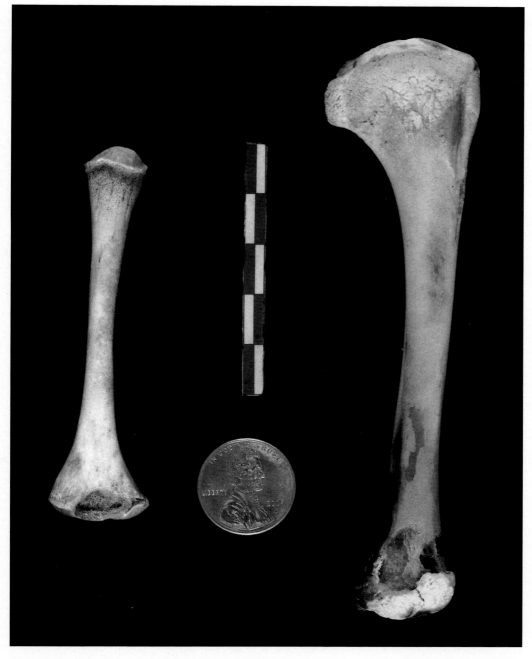

Fig. 15-02. An infant human left humerus (posterior view) is compared to a duck's left humerus (caudal view).

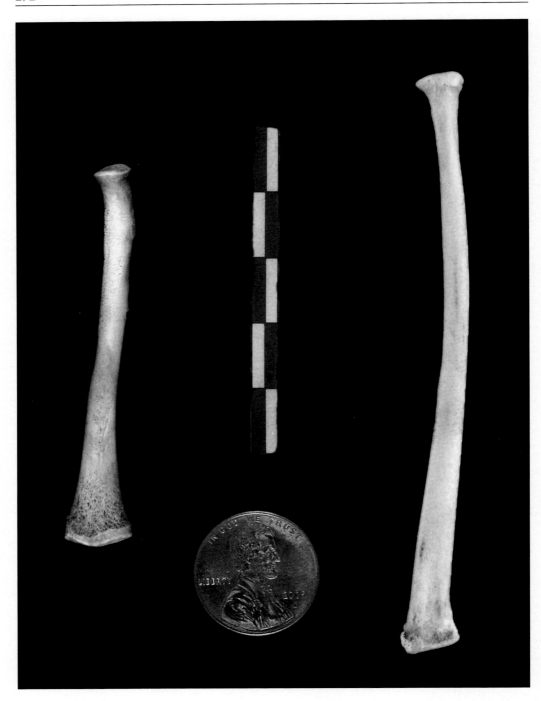

Fig. 15-03. A juvenile human right radius (anterior view) is compared to a duck's right radius (cranial view).

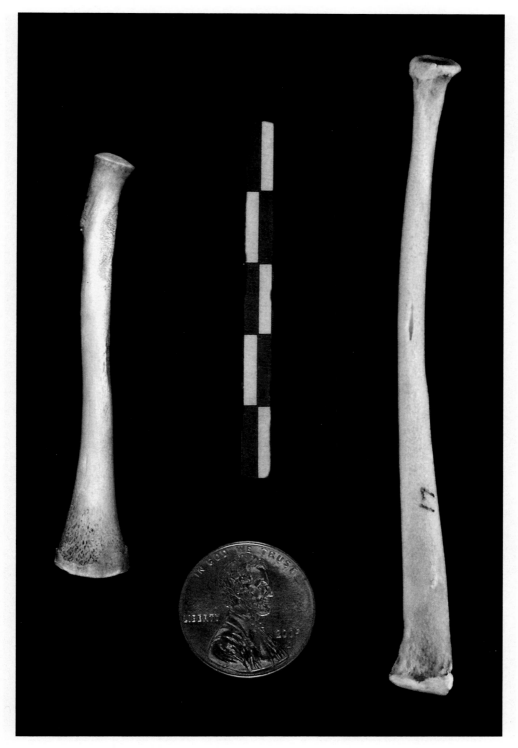

Fig. 15-04. A juvenile human right radius (posterior view) is compared to a duck's right radius (caudal view).

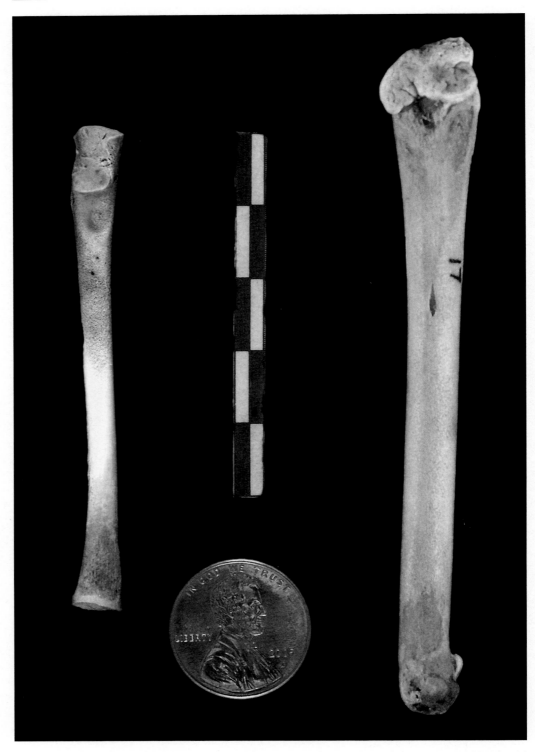

Fig. 15-05. An infant human right ulna (anterior view) is compared with a duck's right ulna (cranial view).

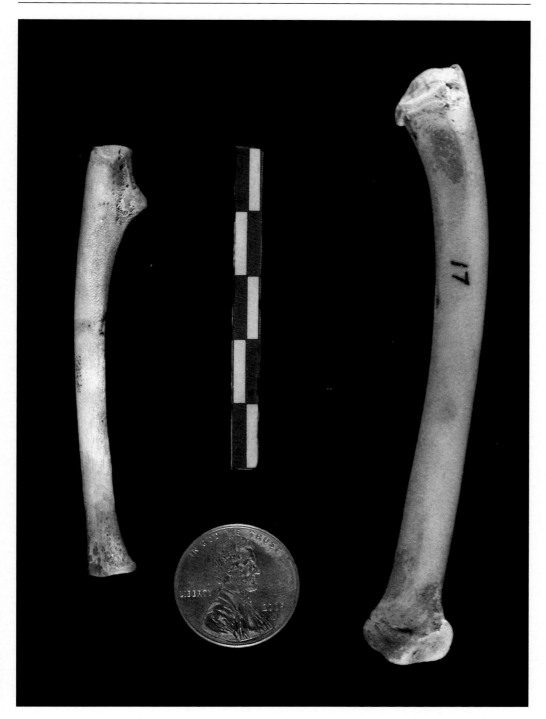

Fig. 15-06. An infant human right ulna (lateral view) is compared to a duck's right ulna (caudal view).

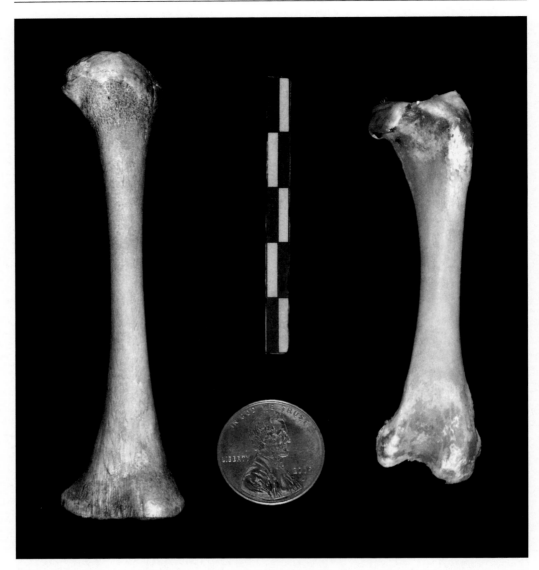

Fig. 15-07. An infant human left femur (anterior view) is compared to a duck's left femur (cranial view).

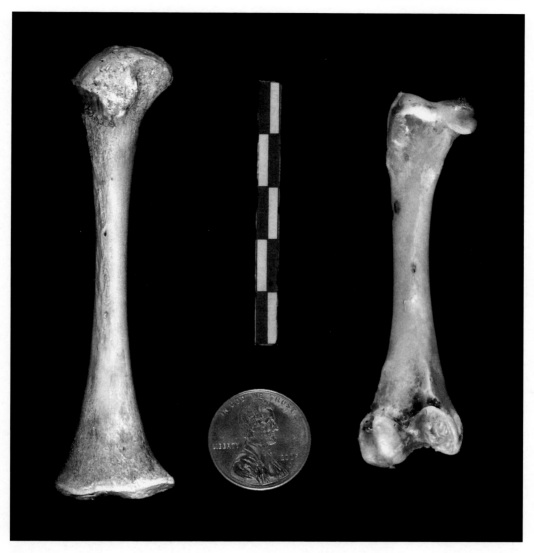

Fig. 15-08. An infant human left femur (posterior view) is compared to a duck's left femur (caudal view).

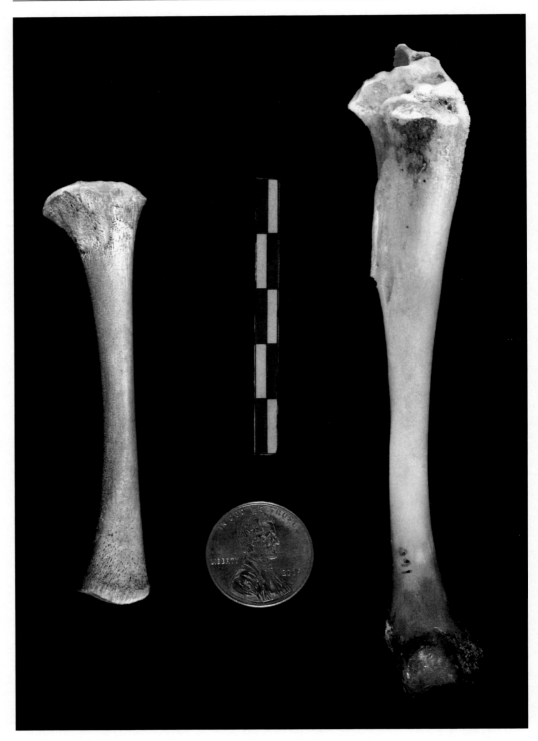

Fig. 15-09. An infant human left tibia (anterior view) is compared to a duck's left tibiotarsus (cranial view).

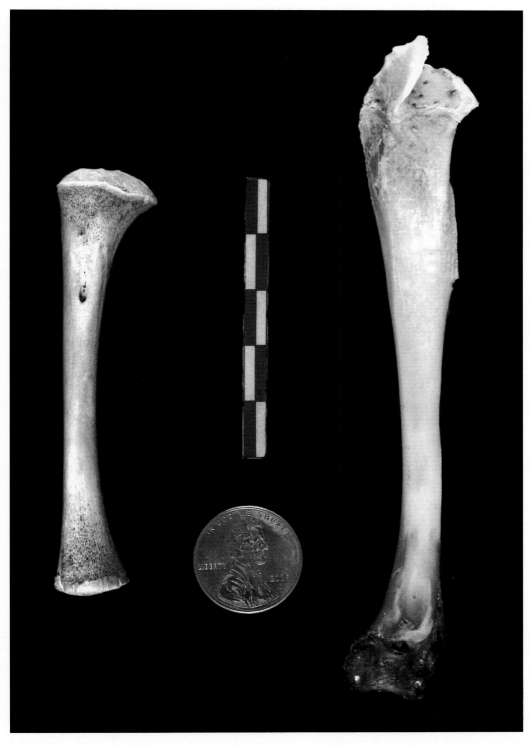

Fig. 15-10. An infant human left tibia (posterior view) is compared to a duck's left tibiotarsus (caudal view).

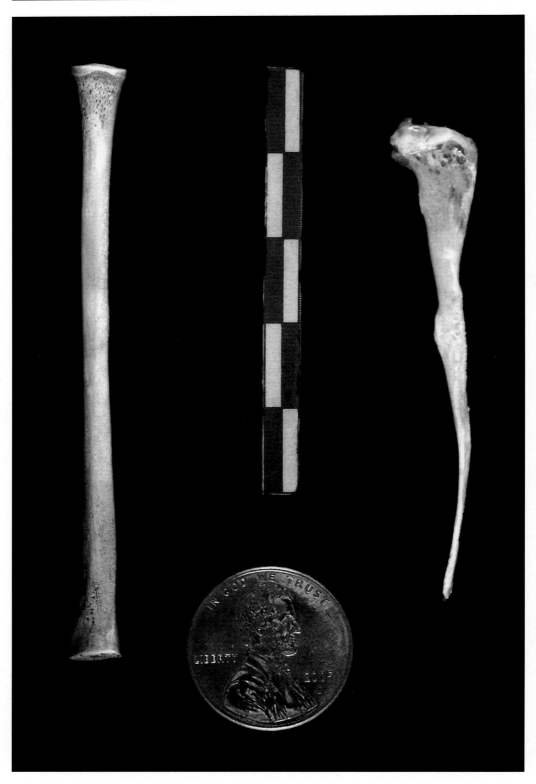

Fig. 15-11. An infant human left fibula (medial view) is compared to a duck's left fibula (medial view).

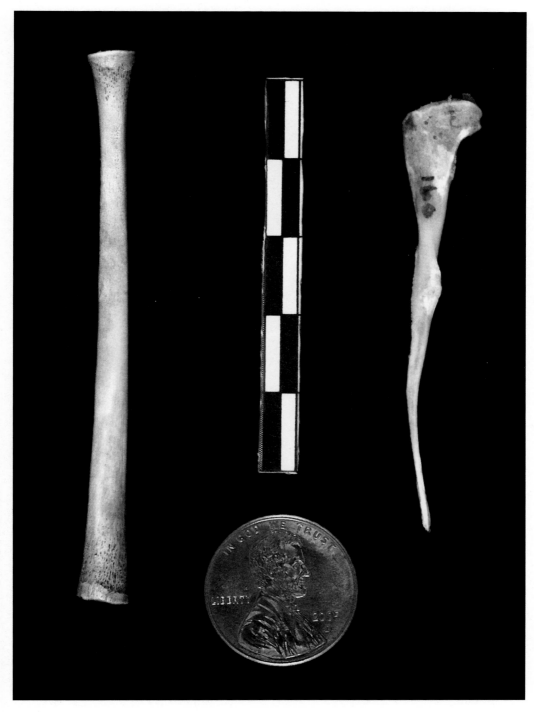

Fig. 15-12. An infant human left fibula (lateral view) is compared to a duck's left fibula (lateral view).

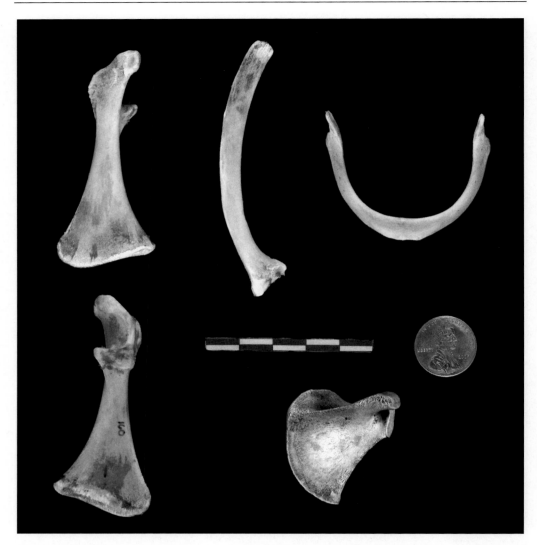

Fig. 15-13. The bird pectoral girdle includes three elements: the coracoid, the scapula, and the furcula (wishbone). A duck's right coracoid (ventral and dorsal views) is pictured on the left top and bottom, a duck's scapula (costal view) is pictured on the top middle, and a duck's furcula is pictured on the top right. The bottom right of the photo shows an infant human right scapula (posterior view).

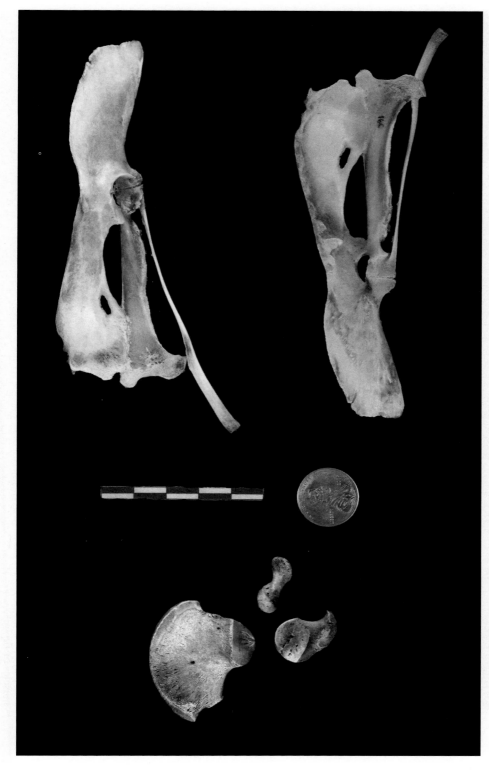

Fig. 15-14. An infant human right innominate (lateral view) is compared to a duck's right pelvis (lateral and medial views). Note that the duck's acetabulum is perforated.

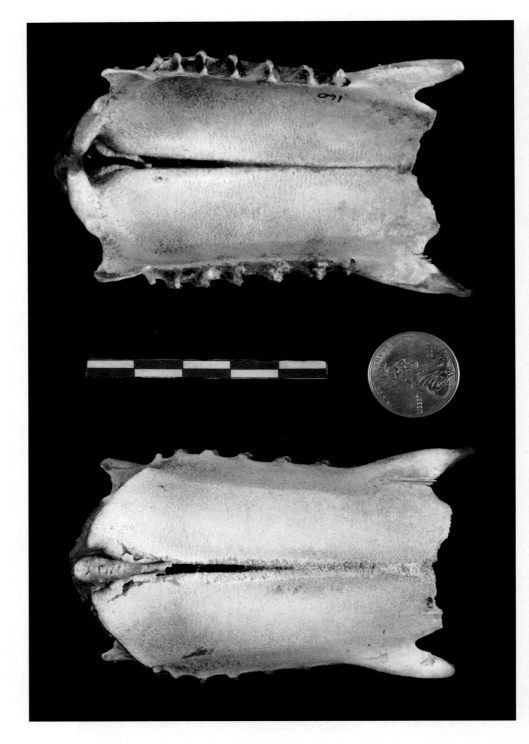

Fig. 15-15. Ventral and dorsal views of the duck's sternum. The keel of the sternum is enlarged and serves as an area of origin for the major muscles of flight.

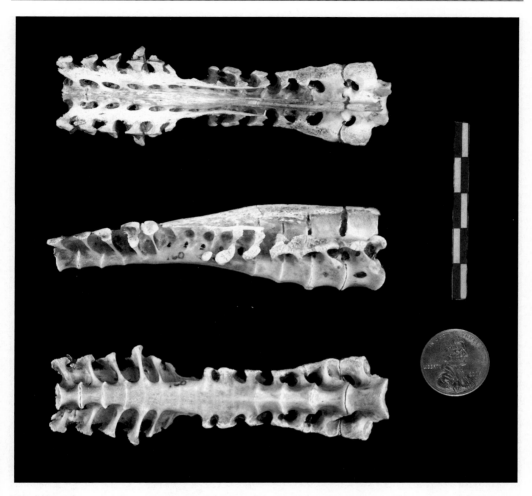

Fig. 15-16. Dorsal, lateral, and ventral views of the duck's synsacrum. In birds, the synsacrum is a made up of the fused lumbar and sacral vertebrae.

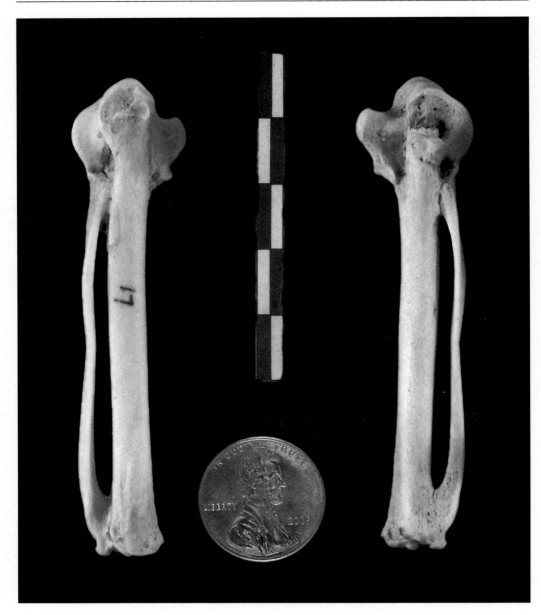

Fig. 15-17. The duck's right carpometacarpus (dorsal and volar views).

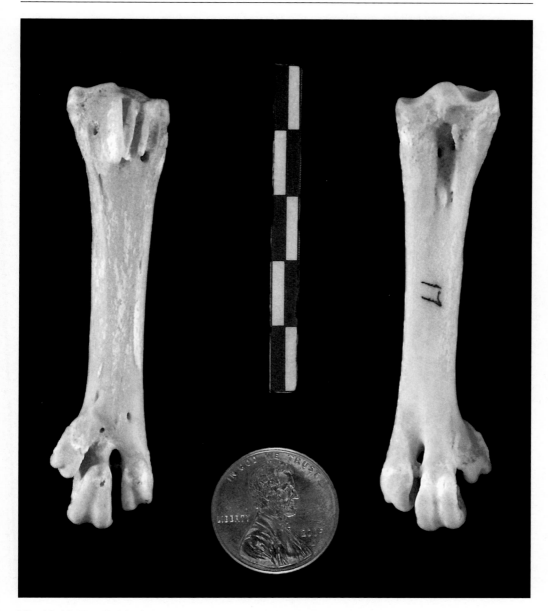

Fig. 15-18. The duck's right tarsometatarsus (dorsal and plantar views). This bone is composed of the fused distal tarsal bones along with three metatarsals.

16 Human vs Chicken

Fig. 16-00. Chicken skull (missing beak).

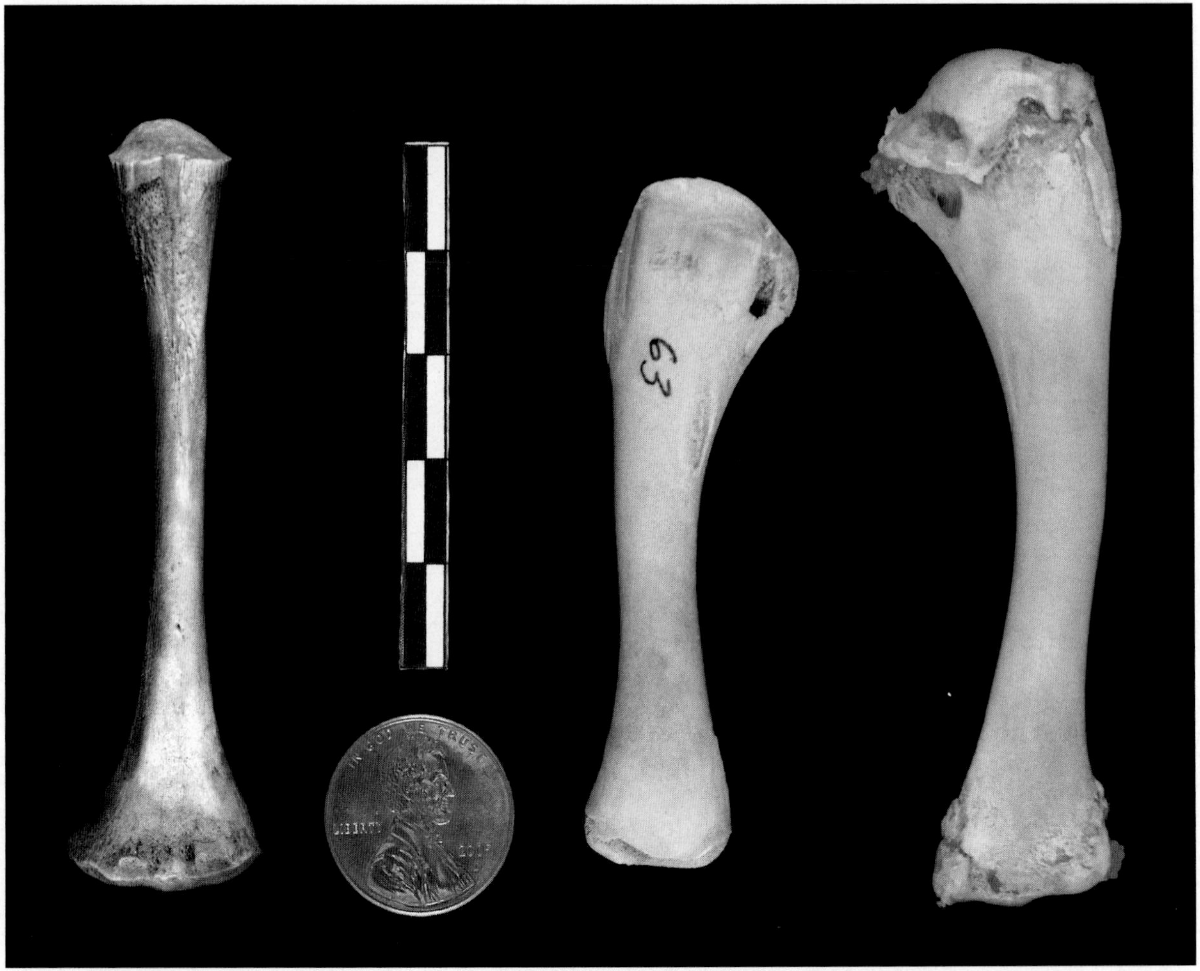

Fig. 16-01. An infant human left humerus (anterior view) is compared to chicken left and right humeri (cranial views).

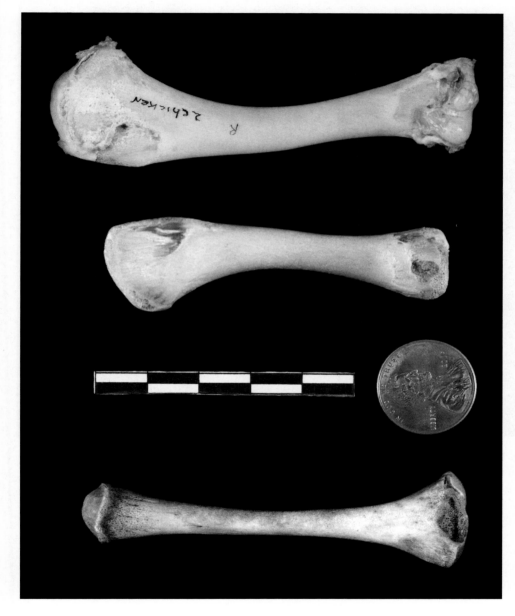

Fig. 16-02. An infant human left humerus (posterior view) is compared to chicken left and right humeri (caudal views).

291

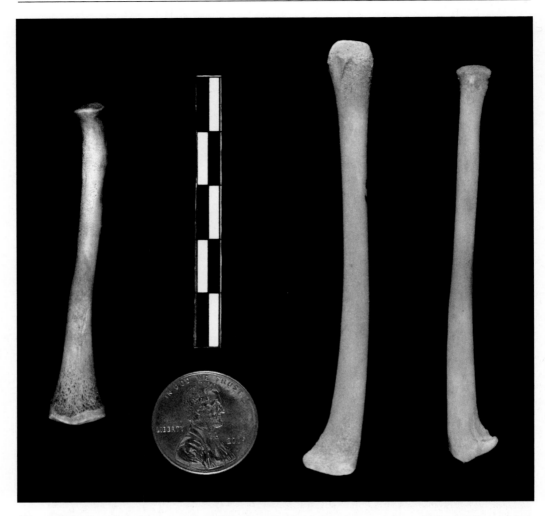

Fig. 16-03. An infant human right radius (anterior view) is compared to chicken right and left radii (cranial views).

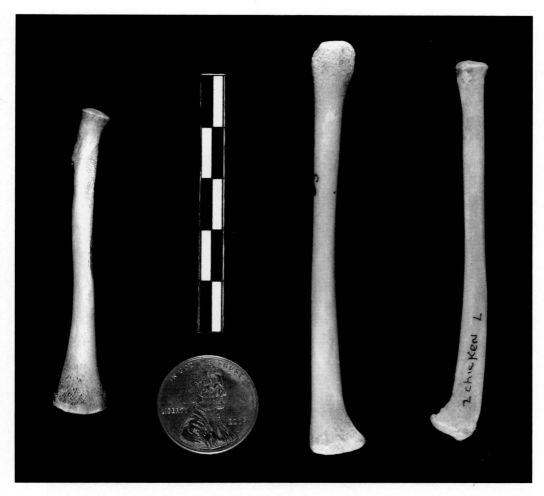

Fig. 16-04. An infant human right radius (posterior view) is compared to chicken right and left radii (caudal views).

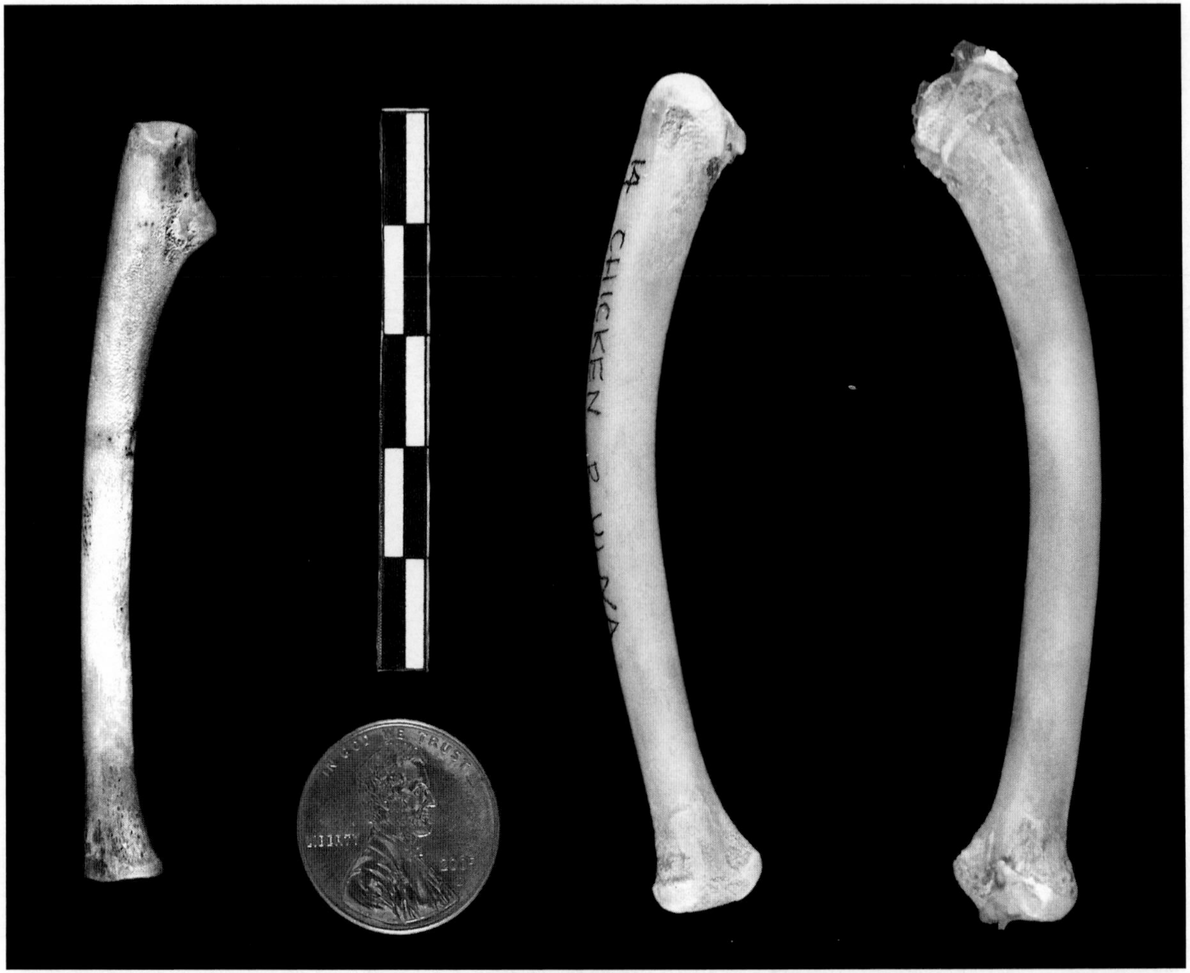

Fig. 16-05. An infant human right ulna (lateral view) is compared to chicken right and left ulnae (lateral views).

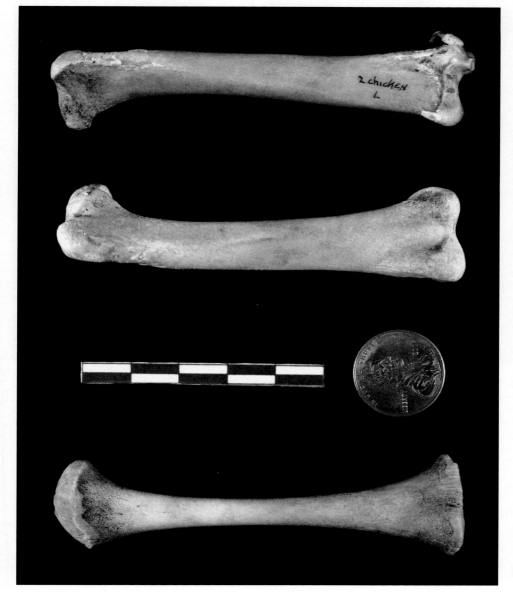

Fig. 16-06. An infant human right femur (anterior view) is compared to chicken right and left femora (cranial views).

295

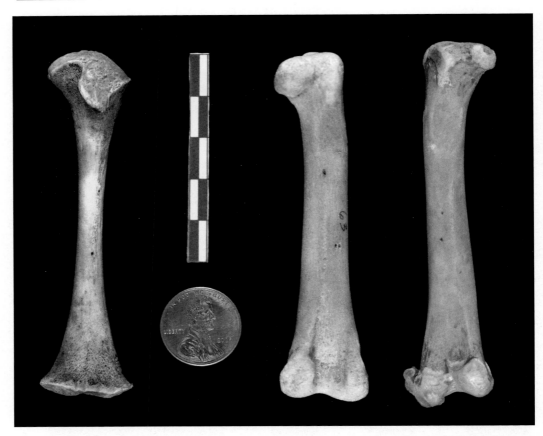

Fig. 16-07. An infant human right femur (posterior view) is compared to chicken right and left femora (caudal views).

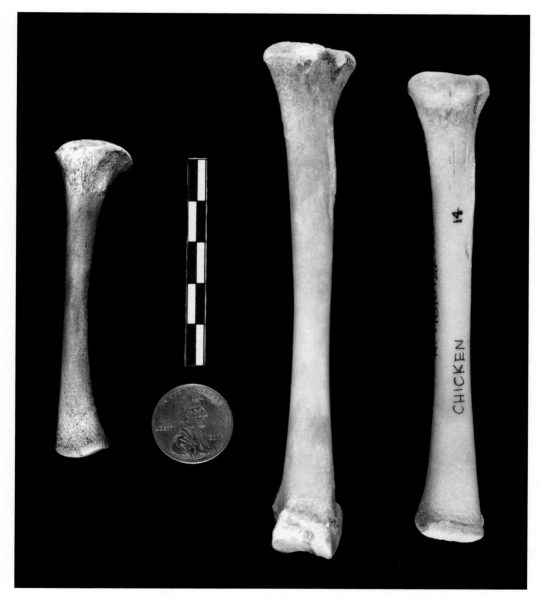

Fig. 16-08. An infant human right tibia (anterior view) is compared to two chicken right tibiotarsi (cranial views).

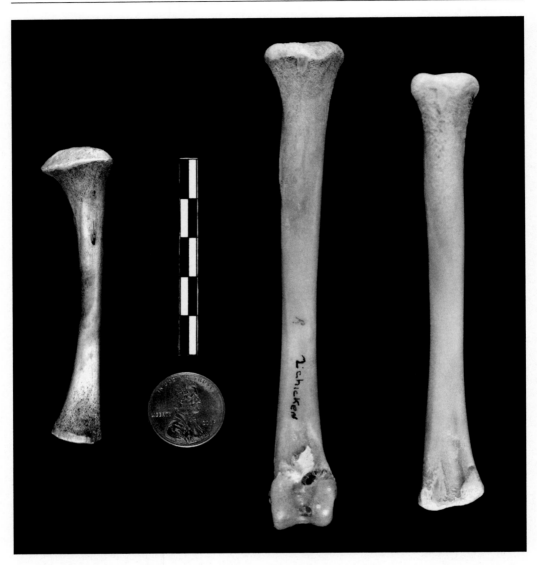

Fig. 16-09. An infant human right tibia (posterior view) is compared to two chicken right tibiotarsi (caudal views).

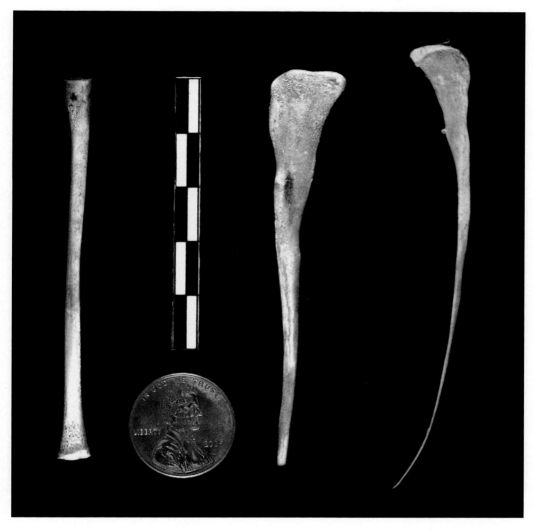

Fig. 16-10. An infant human right fibula (medial view) is compared to chicken right and left fibulae (medial view).

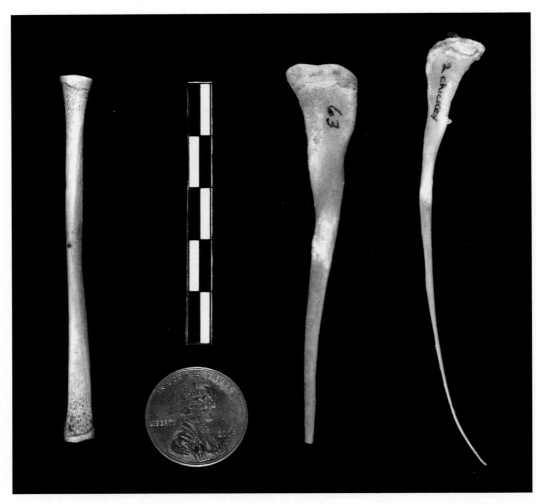

Fig. 16-11. An infant human right fibula (lateral view) is compared to chicken right and left fibulae (lateral view).

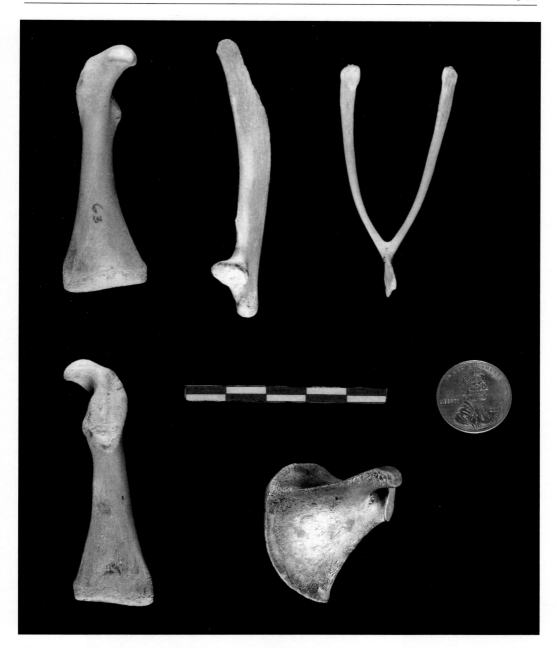

Fig. 16-12. The bird pectoral girdle includes three elements: the coracoid, the scapula, and the furcula (wishbone). A chicken's right coracoid (ventral and dorsal views) is pictured on the left top and bottom, a chicken's scapula (costal view) is pictured on the top middle, and a chicken's furcula is pictured on the top right. The bottom right of the photo shows an infant human right scapula (posterior view).

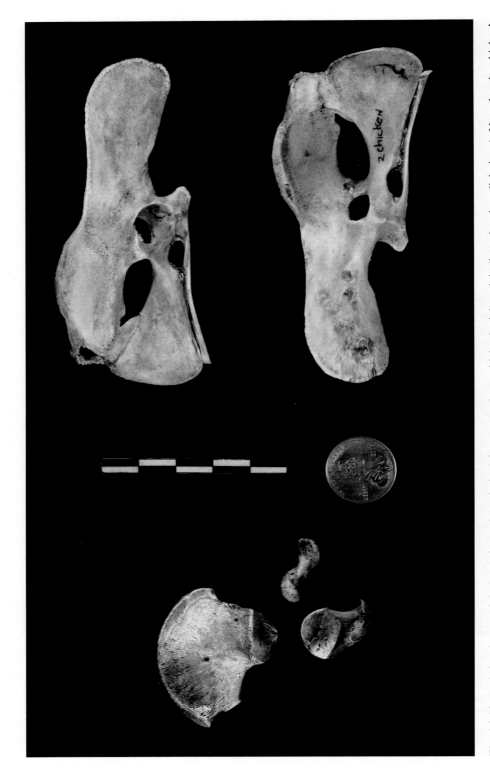

Fig. 16-13. An infant human right innominate (lateral view) is compared to a chicken's right pelvis (lateral and medial views). Note that the chicken's acetabulum is perforated.

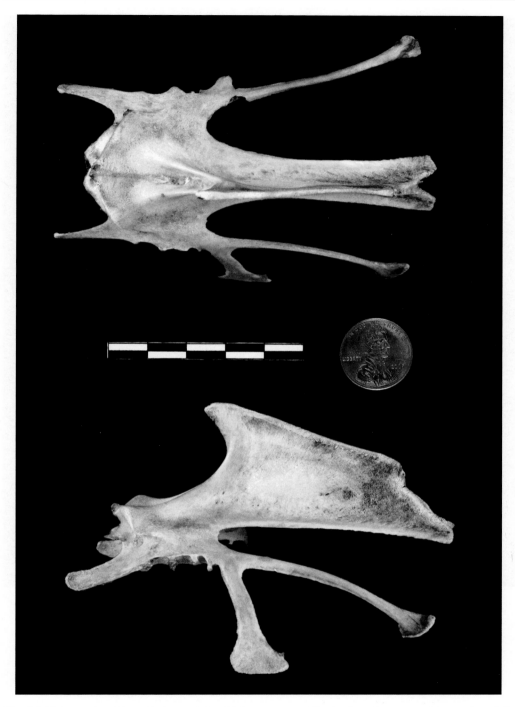

Fig. 16-14. Ventral and lateral views of the chicken sternum. The keel on the sternum forms the origin for the muscles involved in flight.

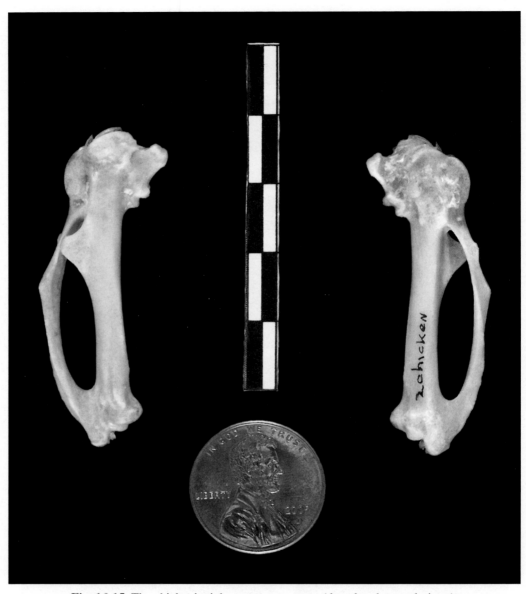

Fig. 16-15. The chicken's right carpometacarpus (dorsal and ventral views).

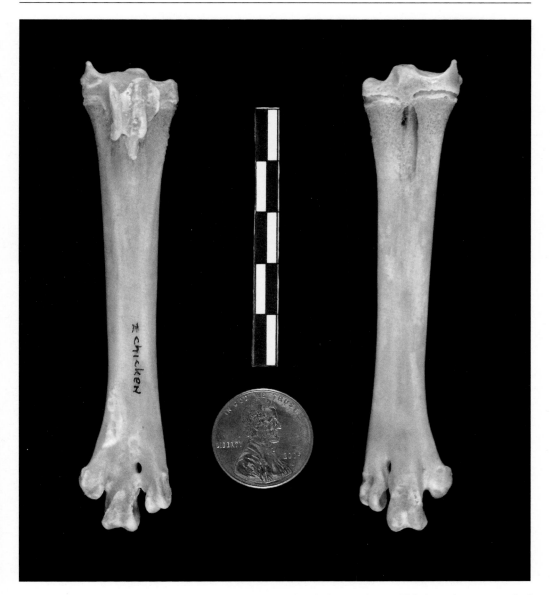

Fig. 16-16. The chicken's right tarsometatarsus (dorsal and plantar views). This bone is composed of the fused distal tarsal bones along with three metatarsals. Male chickens have a bony spur core on the midshaft of the tarsometatarsus.

17 Miscellaneous

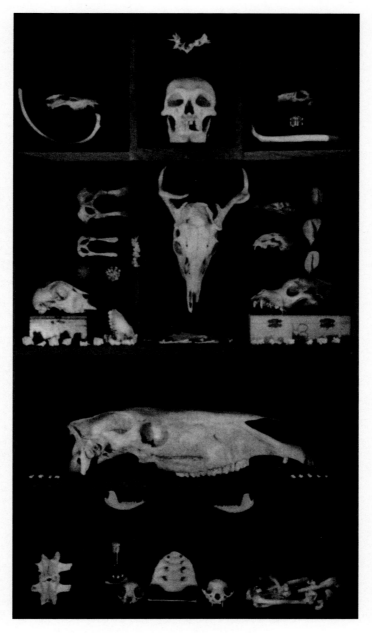

Fig. 17-00. Various animal and human bones in a case.

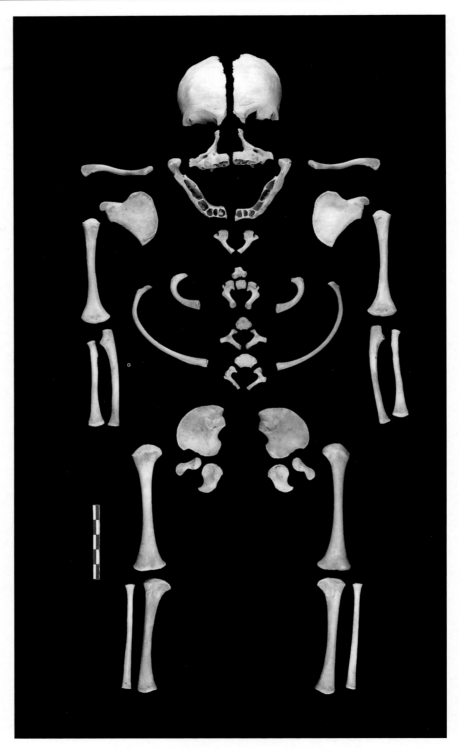

Fig. 17-01. Major skeletal elements of an infant human skeleton positioned in anatomical order (i.e., anterior view with hands at the side and palms forward).

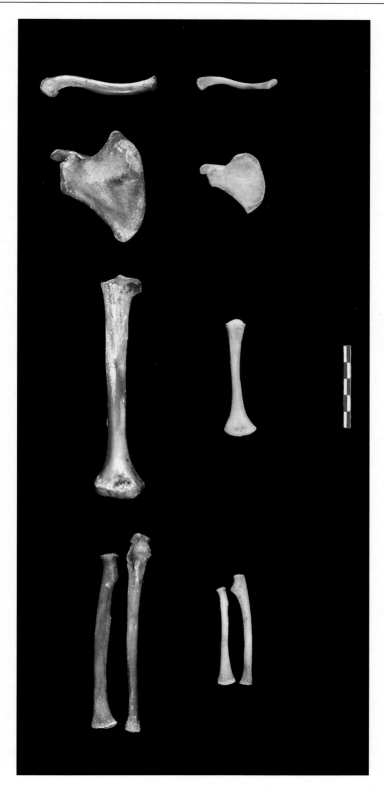

Fig. 17-02. Comparison of right arm and shoulder bones from a toddler (3-4 years old, pictured on left) to corresponding human infant (newborn, pictured on right).

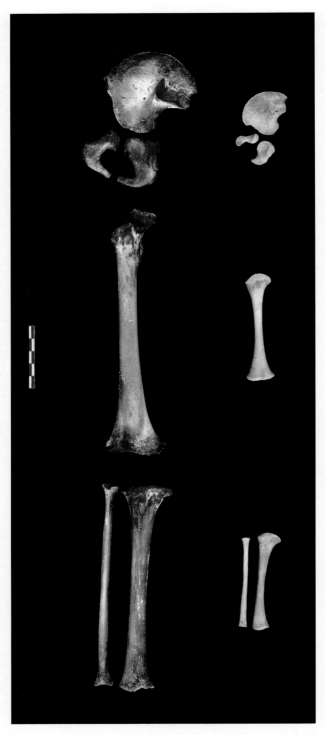

Fig. 17-03. Comparison of right pelvis and leg bones from a toddler (3–4 years old, pictured on left) to corresponding human infant (newborn, pictured on right).

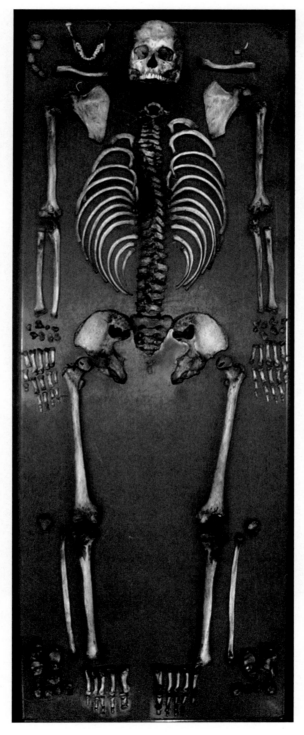

Fig. 17-04. Adult human skeleton positioned in anatomical order (i.e., anterior view with hands at the side and palms forward).

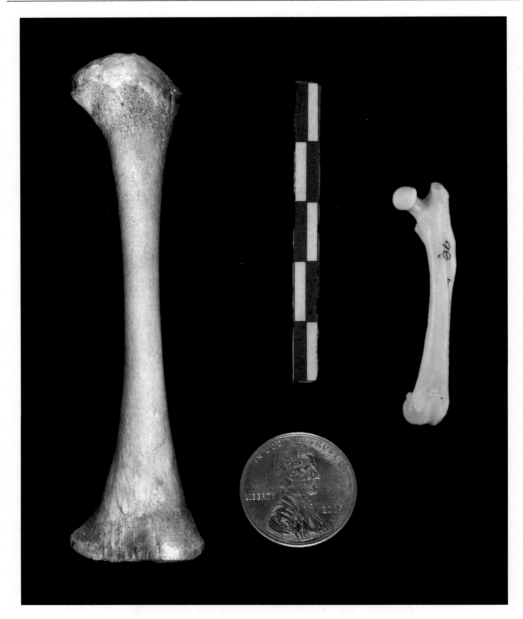

Fig. 17-05. An infant human left femur (anterior view) is compared to the left femur of a rat (cranial view).

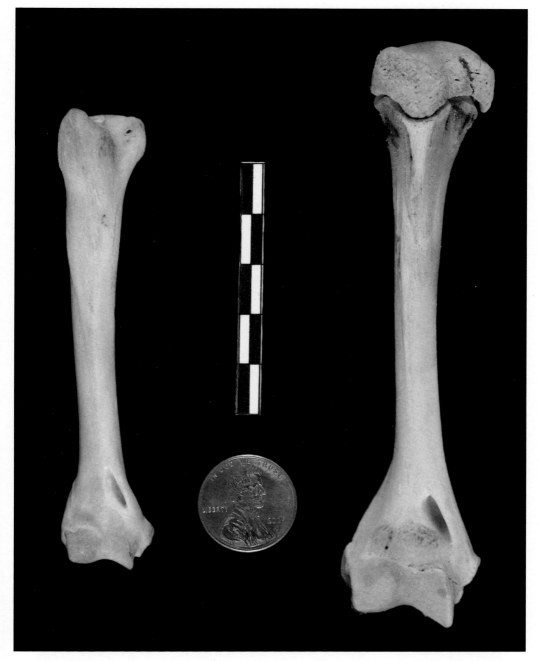

Fig. 17-06. A cat's right humerus (cranial view) is compared to a juvenile bobcat's right humerus (cranial view). Note the presence of the supercondylar foramen on both specimens.

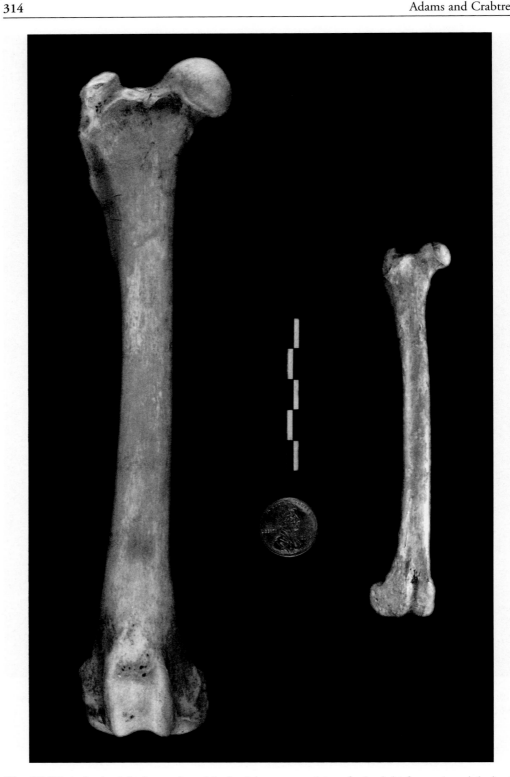

Fig. 17-07. A dog's right femur (cranial view) is compared to a fox's right femur (cranial view). Note the morphological similarities between the two canid species.

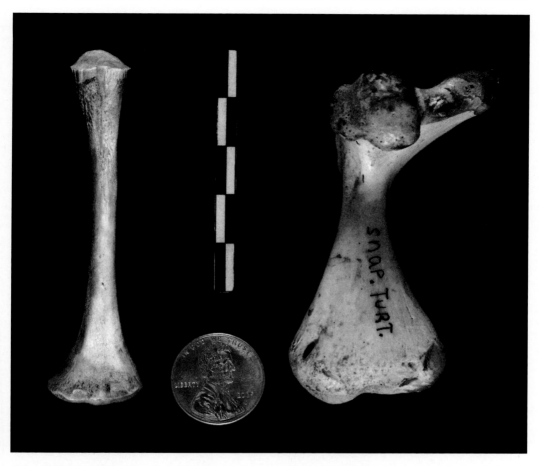

Fig. 17-08. An infant human left humerus (anterior view) is compared to a snapping turtle's left humerus (cranial view).

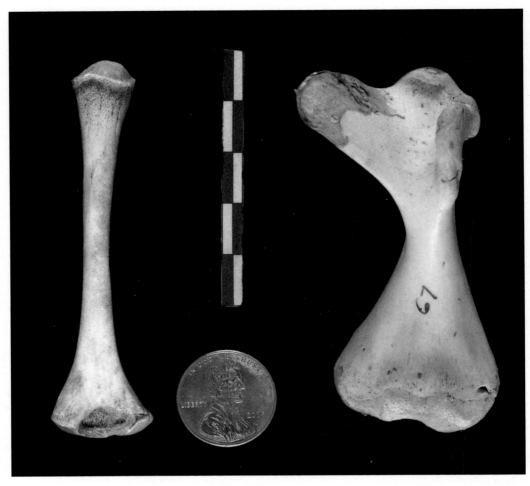

Fig. 17-09. An infant human left humerus (posterior view) is compared to a snapping turtle's left humerus (caudal view).

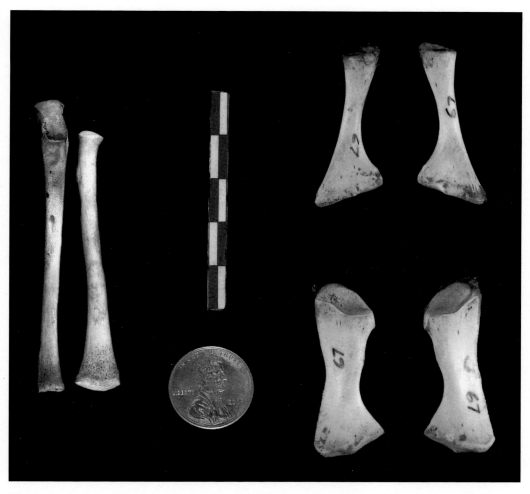

Fig. 17-10. An infant human left radius and ulna (anterior views) are compared to turtle right and left radii and ulnae.

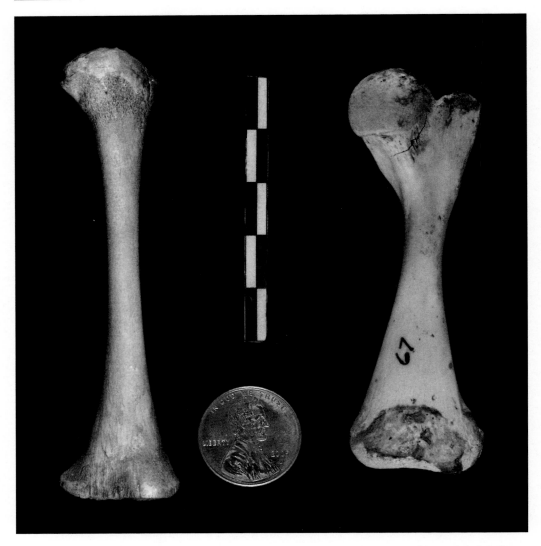

Fig. 17-11. An infant human left femur (anterior view) is compared to a snapping turtle's left femur (cranial view).

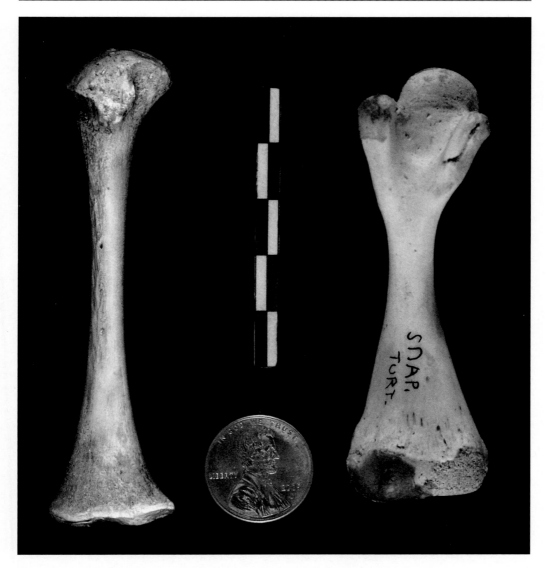

Fig. 17-12. An infant human left femur (posterior view) is compared to a snapping turtle's left femur (caudal view).

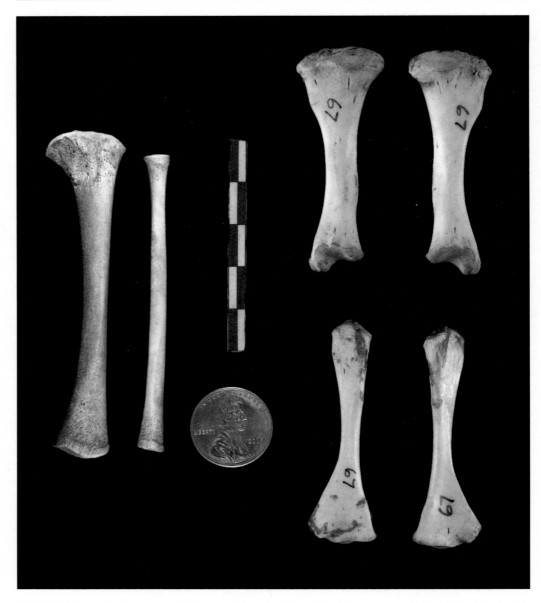

Fig. 17-13. An infant human left tibia and fibula (anterior views) are compared to right and left turtle tibiae and fibulae.

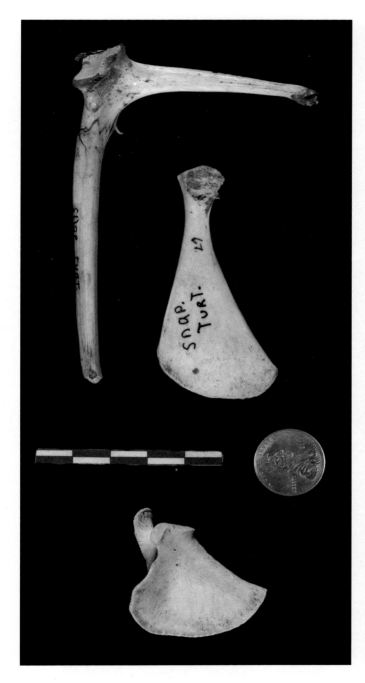

Fig. 17-14. An infant human left scapula (anterior view) is compared to a turtle's left shoulder girdle. Note that the turtle's shoulder girdle includes three elements: the scapula, the acromion, and the anterior coracoid. The scapula and the acromion are fused. The clavicle is fused to the plastron, or base of the shell.

321

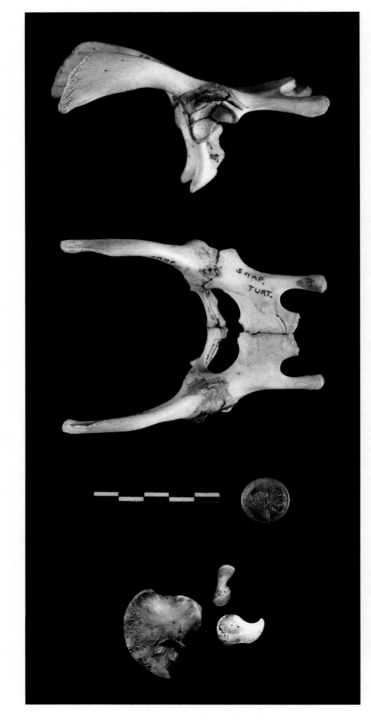

Fig. 17-15. An infant human left innominate (medial view) is compared to a turtle's pelvic girdle (ventral and lateral views).

18 Traces of Butchery and Bone Working

Pam J. Crabtree and Douglas V. Campana

Introduction

Animal bones often reveal marks of butchery associated with meat or marrow processing. Butchered bones may be recovered as important behavioral evidence from archaeological sites, or they may be collected within the forensic context and mistaken for human bones with marks of trauma. Archaeologists can use these butchery marks to study the ways in which past human populations butchered, distributed, and consumed meat. In a forensic context, tool marks on bone may be a good indication that the remains are nonhuman in origin, but this is not always the case. The intentional dismemberment of a human body by another individual (usually with the goal of hindering identification or facilitating transportation of the remains) may mimic the appearance of a butchered cow or pig to the untrained observer. An experienced osteologist should always be consulted if there is any doubt. Figure 18-01 shows several commercially butchered cow bones that were mistaken for human remains and turned over to law enforcement. Figures 18-02 and 18-03 are views of a human femur and humerus from an individual who was murdered and subsequently dismembered with a power saw.

This chapter will illustrate the types of butchery marks that are typically found on animal bones from contemporary, historic, and prehistoric contexts. Butchered remains of cattle and pigs are most commonly found on historic archaeological sites in North America, whereas prehistoric sites often yield substantial numbers of butchered deer bones. Cow and pig are also the most frequently encountered butchered bones from the contemporary context and, based on their size, they are commonly mistaken for human bones. The chapter will also include schematic drawings of a pig, a cow, a lamb, and a deer that show the major cuts of meat. The chapter will conclude with a brief discussion of some of the ways in which bone has been used as a raw material for the manufacture of tools and other artifacts.

Modern Butchery: Eighteenth Century to the Present

Butchers have two main goals: they seek to divide the animal carcass into a number of smaller and more manageable pieces, and, in many cases, they also seek to remove some of the meat from the animal's skeleton. The typical cuts of beef, pork, mutton, and venison are illustrated in Figs. 18-04 through 18-07. Since the 18th century, American butchers have used saws to butcher large animals, and saw marks are some of the most common traces of butchery seen on modern animal skeletons. Hand saws were used in the 18th and 19th centuries, but power saws are most commonly used in modern butchery. Sawing can be readily recognized by the characteristic kerfs, or saw marks, left behind on the sawn bone surface. Hand-sawn bone can be distinguished by the somewhat irregular sawn surface, with groups of parallel, often coarse, striations at angles with one another on the kerf walls; machine-sawn bone usually shows a flat, polished surface with fine, parallel striations on the kerf walls. The analysis of kerf features can also be very informative in the forensic context when assessing tool marks left on bone from human dismemberment cases (Symes, et al. 1998; Symes, et al. 2002). For the processing of food remains, saws are used both to split the carcass into sides and to further subdivide the carcass into joints of meat. The following are some typical examples of saw marks on butchered bones from modern contexts between 18th- and 19th-century archaeological sites.

Figure 18-08 shows a cow scapula that has been sawn with a hand saw. This bone dates to the 18th century and were recovered from deposits associated with the New York City poor house, located on the grounds of the modern City Hall. Figure 18-09 shows a cow's atlas (C1) that has been sagitally sawn. This example comes from a forensic case file from New Jersey. Figures 18-10 and 18-11 are sawn bones from a late 20th-century pig farm in central New Jersey. In this case, a saw has been used to separate the lower limbs from the meatier portions of the upper fore- and hindlimbs. The central portion of the tibial shaft (Fig. 18-10) and the distal parts of the radial and ulnar shafts (Fig. 18-11) have all been sawn through.

Saws are commonly used to divide an animal's carcass into a series of joints of meat. Figure 18-12 shows a cow's humerus from a 19th-century deposit at the archaeological site of Fort Johns in Sussex County, New Jersey (Crabtree, et al. 2002). The humerus has saw marks near the distal end and on the central portion of the shaft. A second saw mark is visible on the central portion of the shaft where the butcher's saw must have slipped early in the process. These saw marks were used to produce a chuck or arm roast.

Saws are often used to divide beef ribs into small, 4- to 6-in. (10–15 cm) sections. Figure 18-13 (top) shows a central portion of a cow's rib with saw marks on either end, producing a cross-rib cut. This example was also recovered from a 19th or early 20th-century context at the Fort Johns site. A second rib section from Fort John's (Figure 18-13, bottom) includes the articular portion of the bone (the part closest to the vertebral column).

Other typical examples of 19th-century butchery from the Fort Johns site include a cow's ilium (Fig. 18-14, left) that has been sawn into a roughly 1-in. (2.5 cm) segment, probably for a steak, and the glenoid portion of a cow's scapula (Figure 18-14, second from left) that has been sawn through, possibly to produce a chuck roast. The Fort Johns excavations also produced a dorsal spine of a cow's thoracic vertebra (Fig. 18-14, second from right) that has been sawn through. Large, sawn sections of less meaty elements, such as tibial and radial shafts of cattle, are often used as soup bones (Milne and Crabtree 2002: 164). Figure 18-14 (right) shows a sawn section of a cow's tibia from the Five Points site, a 19th-century multiethnic neighborhood in lower Manhattan (Milne and Crabtree 2001).

Knife cuts and other butchery traces can often times be seen on modern faunal remains. For example, Figure 18-15 shows an immature pig tibia that has been made into a spiral ham. The circular cut marks can be seen on the proximal shaft of the tibia.

Butchery Using Cleavers and Heavy Knives

Earlier than about 1700 A.D., most butchery was carried out using cleavers and heavy knives. These tools were also commonly used for home butchery during the 18th and 19th centuries. Figure 18-16 shows a cow proximal femur that has been split using a cleaver. An initial, unsuccessful chop mark can also be seen on the proximal end near the femoral head. This example comes from the Middle Saxon (~650-850 A.D.) settlement site of Brandon in eastern England (Carr, et al. 1988).

Lower limb bones, especially the metacarpals and metatarsals of the ruminant artiodactyls, are often split for the extraction of marrow. Figure 18-17 shows a cattle metatarsal that has been axially split. This example was also recovered from the Brandon site.

Professional butchers can dismember a large animal carcass skillfully using sharp cleavers. The following examples from the Roman (2nd to 4th century A.D.) site of Icklingham in eastern England show how Roman butchers could dismember animal carcasses using sharp iron tools. Figure 18-18 shows the types of chop marks that are typically produced by a heavy knife or cleaver. These marks appear near the gonial angle of a cow's mandible. Figure 18-19 shows similar heavy chop marks on the caudal portion of a cow radius. Figure 18-20 shows a cow's pubis that has been chopped through using a cleaver.

Prehistoric Butchery

Before the development of high quality metal tools, butchery was carried out using stone tools. The earliest examples of animal bones butchered with stone tools date back to about 2.6 million-yr-ago in East Africa (Semaw 2000; Semaw, et al. 2003). In North America, animal bones were butchered using stone tools until the early 17th century when metal knives and other tools were introduced from Europe. For example, a well-preserved iron cleaver was recovered from the 17th-century site of Martin's Hundred in Virginia (Noël Hume 1979: 147). Archaeologists working on prehistoric sites in the Americas are likely to encounter animal bones that have been butchered using stone knives and flakes. When compared with the traces left by iron tools, butchery traces left by stone tools are far more subtle. For example, a butcher using iron tools will often separate the femur from the acetabulum by chopping through the neck of the femur (*see* Figure 18-16). A prehistoric butcher using stone tools will cut around the outside of the acetabulum in order to detach, or disarticulate, the femur.

Figure 18-21 shows an astragalus of a wild goat *(Capra aegagrus)* that was butchered by the Neanderthals inhabiting Shanidar Cave in Iraq between 60,000 and 44,000 BP (Solecki 1971). The relatively deep, V-shaped cuts across the bone were left by the edge of a flint flake during the disarticulation of the animal's leg. Figure 18-22 is the phalanx of a Shanidar goat; these clear, somewhat shallower V-shaped cuts were left during the process of skinning the animal. Figure 18-23 is a phalanx of a modern white-tailed deer *(Odocoileus virginianus)* that was experimentally skinned using flint tools similar to those found at Shanidar. The placement and appearance of these markings is very similar to those found on the archaeological specimens.

Bone as a Raw Material

Bone has served as an important raw material throughout nearly all of human history. Bone was commonly used as a raw material for tools and other artifacts until its replacement by plastic after World War II. For example, bone toothbrushes are commonly found on 19th-century archaeological sites. Bone spools were used to hold thread for tatting (lace-making) as recently as the 1930s.

Not all bones are equally suitable for the manufacture of bone tools and artifacts. Prehistoric and historic craftworkers often chose to use ungulate metacarpals, metatarsals, and tibias because these bones are characterized by a relatively long, cylindrical shaft made up of thick, compact bone. This bone is made up of a calcitic matrix deposited around central nutrient canal, forming the osteons, or Haversian system. The osteons of long bones are oriented parallel to the shaft, splaying outwards toward the ends. The combination of thick, compact bone with parallel osteons results in a set of physical characteristics of the material in the bone shaft that is advantageous for the manufacture of bone implements. The shaft of a long bone is markedly anisotropic in strength; that is, it is much stronger when stressed along the length of the bone than it is around the circumference of the bone. Consequently, long bones tend to fracture in long, thin, fragments that are suitable for making into tools and other artifacts.

Figure 18-24 shows a section of a cow's metatarsal that has been sawn at both the proximal and the distal ends. The metatarsus was recovered from the Iron Age (~200 B.C. to 100 A.D.) site of Dún Ailinne in Ireland (Wailes 1990; Wailes 2004). Although sawing was not used in butchery in Ireland until about 1700, saws were used in bone working from the early Iron Age onward. This bone object is the earliest known example of sawing in Ireland (Raftery 1994: 119). This bone appears to be a blank that was prepared for bone working. Figure 18-25 shows a finished bone tool made on a split gazelle metapodial from the 11,000-yr-old site of Salibiya I in the lower Jordan Valley, West Bank (Crabtree, et al. 1991).

Deer antler was also a valuable raw material that was commonly used for handles and other artifacts. Figure 18-26 shows a red deer *(Cervus elaphus)* skull from the medieval Brandon site in eastern England. Both a saw and a heavy cleaver have been used to remove the antlers from the deer's skull. In the Middle Ages, antler was commonly used to manufacture bone combs. Figure 18-27 is an example of an early medieval antler comb from Iceland.

Fig. 18-01. Various examples of cattle bones that were butchered with a saw. These are contemporary and were mistaken for human remains and turned over to the Medical Examiner's Office in New York City. From left to right these include: 1. A cattle distal humerus. The humerus has been sawn through just above the distal end. 2. A cattle thoracic vertebra (caudal view). The vertebra has been sawn in half axially, and the thoracic spine has been sawn off. 3. The head of a cow's rib that has been sawn through. 4. An immature cow's femur (cranial view, note that the head of the femur is unfused). The proximal end of the femur has been sawn through, separating the greater trochanter from the rest of the bone. The shaft of the femur has been sawn through as well. 5. The epiphysis of a distal right femur that has been sawn in half.

327

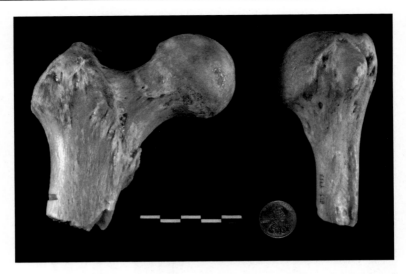

Fig. 18-02. Proximal right human femur and proximal left human humerus (anterior views). These bones are from an individual that was murdered and then dismembered with a power circular saw.

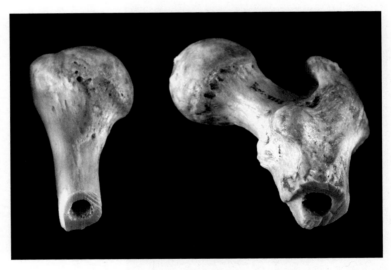

Fig. 18-03. Different views of the dismembered human femur and humerus that show the sawn margins.

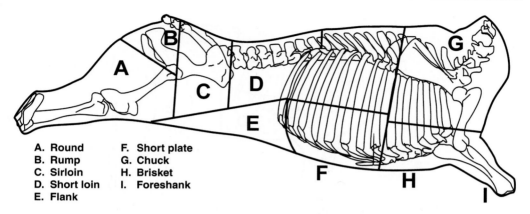

A. Round   F. Short plate
B. Rump   G. Chuck
C. Sirloin   H. Brisket
D. Short loin   I. Foreshank
E. Flank

Fig. 18-04. Modern beef cuts.

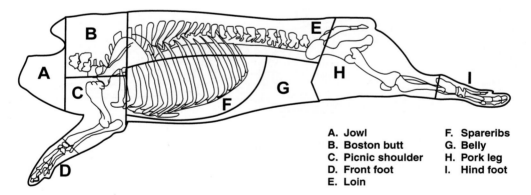

A. Jowl   F. Spareribs
B. Boston butt   G. Belly
C. Picnic shoulder   H. Pork leg
D. Front foot   I. Hind foot
E. Loin

Fig. 18-05. Modern pork cuts.

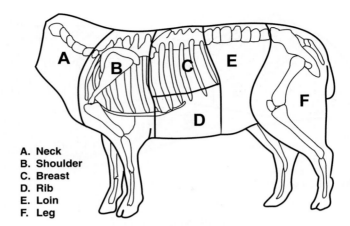

A. Neck
B. Shoulder
C. Breast
D. Rib
E. Loin
F. Leg

Fig. 18-06. Modern lamb (mutton).

A. Neck
B. Shoulder
C. Fore shank
D. Rib
E. Breast
F. Loin
G. Flank
H. Sirloin
I. Leg
J. Hind shank

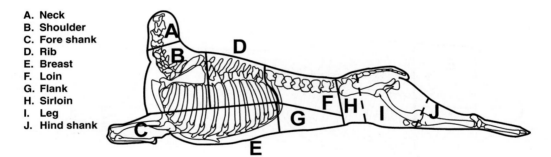

Fig. 18-07. Modern deer (venison) cuts.

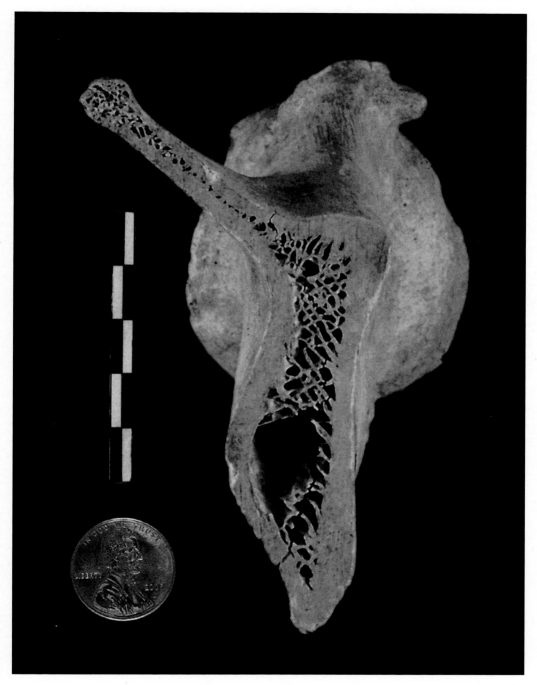

Fig. 18-08. A cow's scapula that has been sawn in half. This example is from the 18th century poor house in New York City.

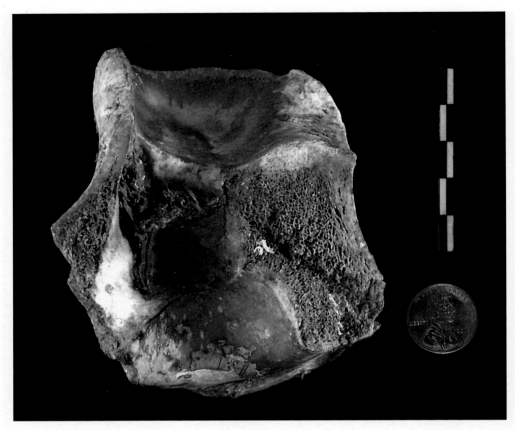

Fig. 18-09. Saw marks can be seen on a cow's cervical vertebra. This bone was part of a forensic case file.

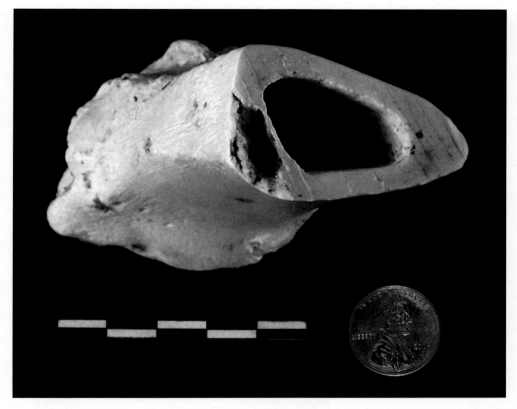

Fig. 18-10. The traces of machine sawing can be seen on this shaft of a pig's tibia from a 20th-century farm in New Jersey.

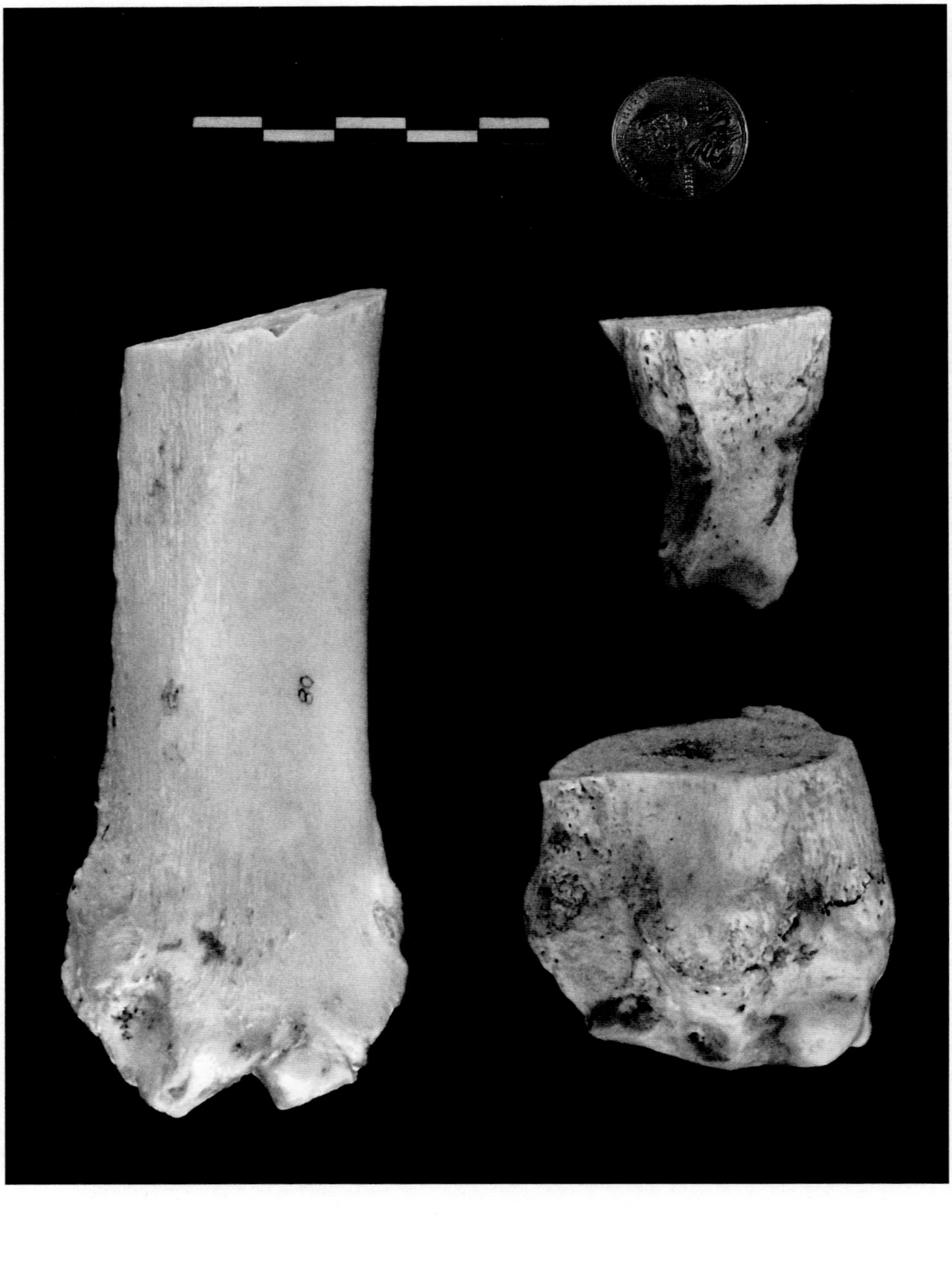

Fig. 18-11. Sawn shafts of a pig's tibia, radius, and ulna from a late 20th-century farm in New Jersey.

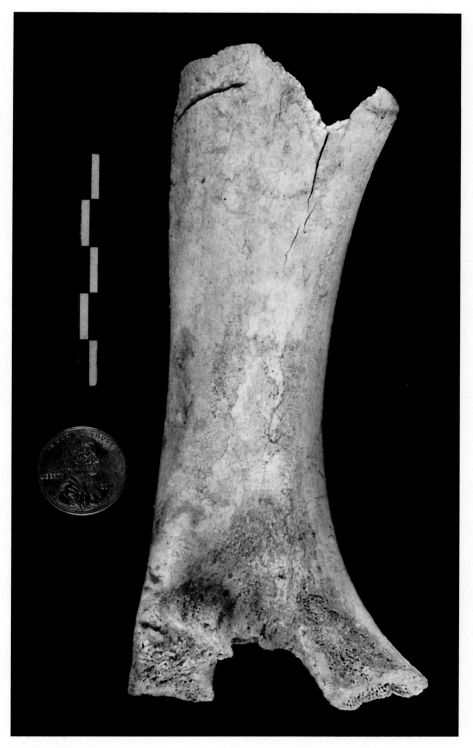

Fig. 18-12. Sawn cow's humerus from the historic Fort Johns site in Sussex County, New Jersey.

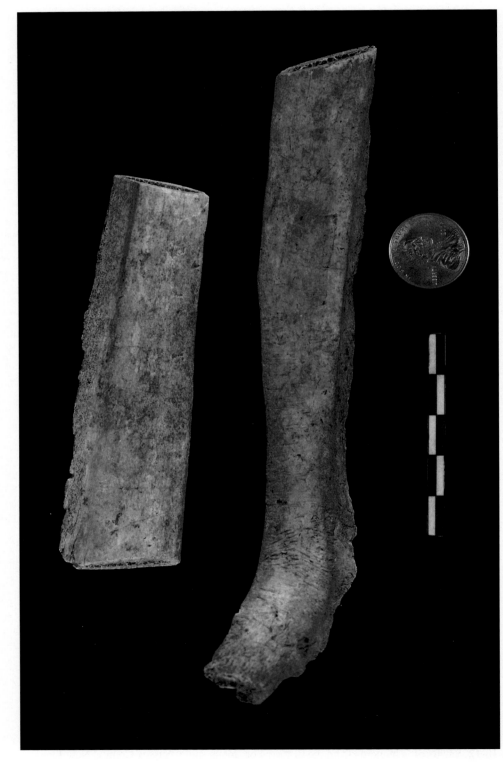

Fig. 18-13 (top). A sawn section of a cow's rib from the Fort Johns site. 18-13 (bottom) A sawn section of a cow's rib including the dorsal portion that articulates with the thoracic vertebrae.

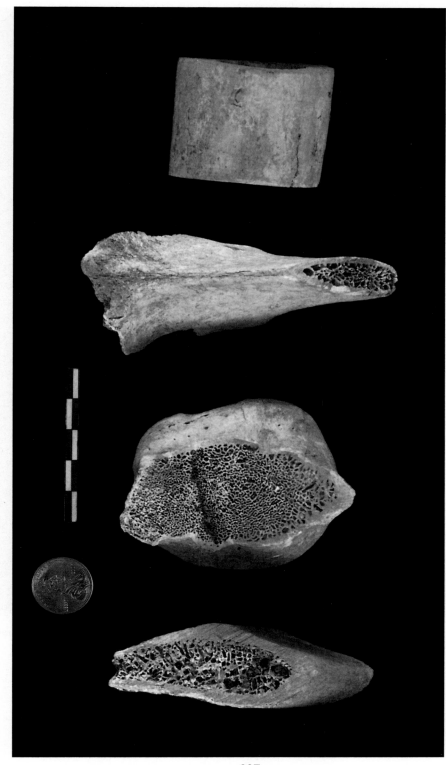

Fig. 18-14. Typical examples of butchery from 19th-century archaeological sites in the northeastern United States. From left to right they include a cow's sawn ilium, scapula, and thoracic spine (from the Fort Johns site in New Jersey), and a section of a cow's tibia (from the Five Points site in Manhattan).

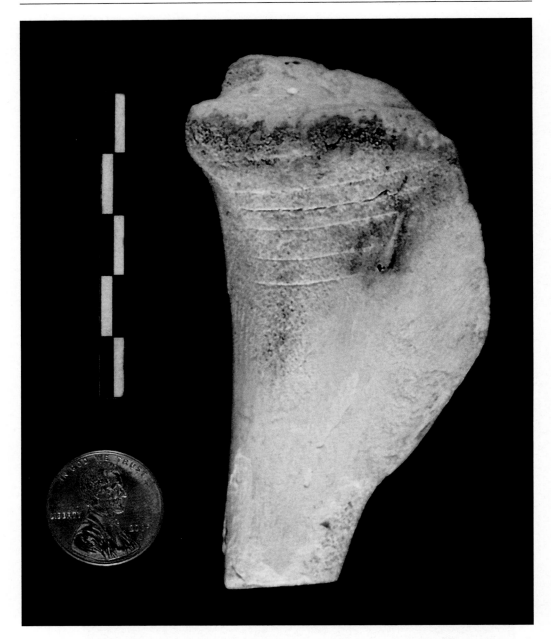

Fig. 18-15. Proximal tibia of a modern pig showing cut marks. This bone is the remnant of a spiral-cut ham.

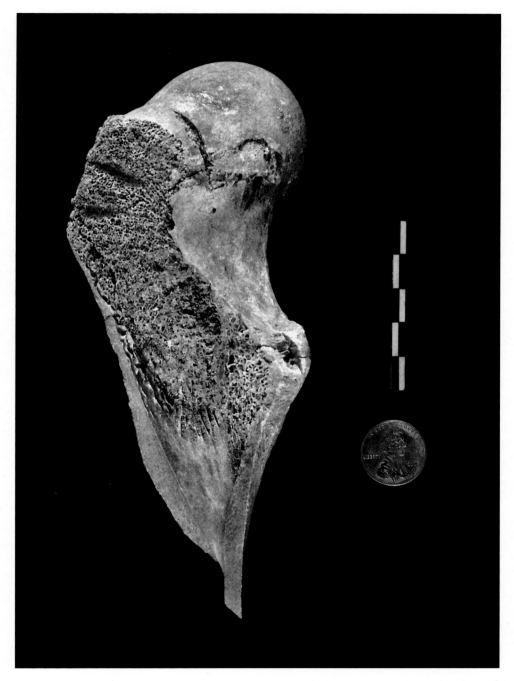

Fig. 18-16. A butchered cow's femur from Brandon, a Middle Saxon (ca. 650-850 AD) site in eastern England.

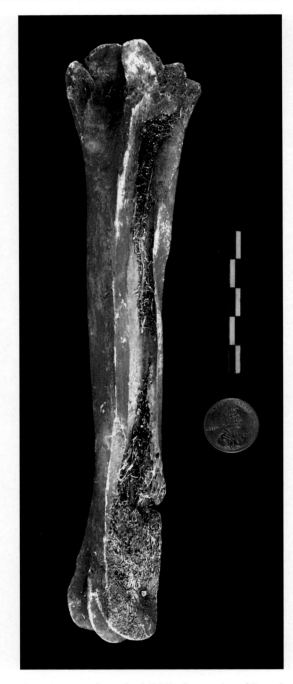

Fig. 18-17. A split cattle metatarsus from the Middle Saxon site of Brandon in eastern England.

Fig. 18-18. Chop marks on a cow's mandible from the 2nd-4th century Roman site of Icklingham in eastern England.

Fig. 18-19. Chop marks on the caudal portion of a cow's radius from Icklingham.

Fig. 18-20. Chop marks on a cow's pubis from Icklingham.

Fig. 18-21. Astragalus fragment of a wild goat *(Capra aegagrus)* from the Middle Paleolithic levels of Shanidar Cave showing stone tool cut marks.

Fig. 18-22. Wild goat *(Capra aegagrus)* second phalanx from the Middle Paleolithic levels of Shanidar Cave in Iraq showing cut marks made with a stone tool.

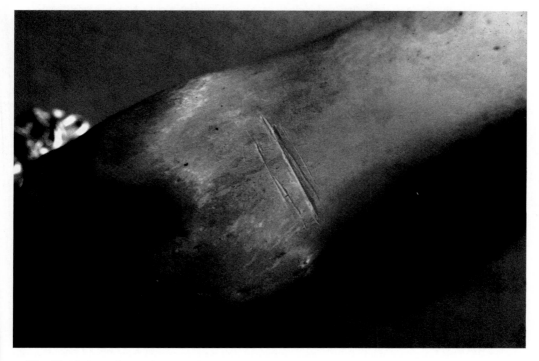

Fig. 18-23. Experimental cut marks made with a stone tool on a white tailed deer first phalanx.

Fig. 18-24. A sawn section of a cow's metatarsal from the Iron Age site of Dún Ailinne in Ireland. This appears to be a blank for bone working.

Fig. 18-25. This bone tool made from a gazelle metapodial was recovered from the 11,000-year-old site of Salibiya I in the West Bank.

Fig. 18-26. A red deer (Cervus elaphus) skull from the Anglo-Saxon site of Brandon in eastern England. One antler has been chopped off, while the other has been sawn off.

Fig. 18-27. An anlter comb from medieval Iceland.

REFERENCES

Bass WM. Human Osteology. 5th ed. Missouri Archaeological Society, Columbia, MO, 2005.

Boessneck J. Osteological differences between sheep (*Ovis aries* Linne) and goat (*Capra hircus* Linne). In: Brothwell D, Higgs E, eds. Science in Archaeology, Praeger Publishers, New York, 1969, pp. 331–358.

Boessneck J, Muller HH, Teichert M. Osteologische Unterscheidungsmerkmale zwischen Schaf (*Ovis aries* Linne) und Ziege (*Capra hircus* Linne). Kuhn-Arch 1964;78:1–129.

Brothwell DR. Digging Up Bones. 3rd ed. Cornell University Press, Ithaca, NY, 1981.

Brown CL, Gustafson CE. A Key to Postcranial Skeletal Remains of Cattle/Bison, Elk, and Horse. Washington State University Reports of Investigations 57. Washington State University, Pullman, WA, 1979.

Byers SN. Introduction to Forensic Anthropology. 2nd ed. Pearson Education, Inc., Boston, 2005.

Carr RD, Tester A, Murphy P. The Middle-Saxon Settlement at Staunch Meadow, Brandon. Antiquity 1988;62:371–377.

Chang KC. The Archaeology of Ancient China. 4th ed. Yale University Press, New Haven, CT, 1986.

Cohen A, Serjeantson D. A Manual for the Identification of Bird Bones from Archaeological Sites. Revised ed. Archetype Publications, London, 1996.

Cornwall IW. Bones for the Archeologist. Macmillan Press, New York, 1956.

Crabtree P, Campana DV, Wright JR. Exploring the Archaeological Potential of French and Indian War Fortifications. CRM 2002;25(3):21,22.

Crabtree PJ, Campana DV, Belfer-Cohen A, Bar-Yosef D. First Results of the Excavations at Salibiya I, Lower Jordan Valley. In: Bar-Yosef O, Valla F, eds. The Natufian Culture of the Levant, International Monographs in Prehistory, Ann Arbor, MI, 1991, pp. 161–172.

Evans HE, de Lahunta A. Miller's Guide to the Dissection of the Dog. WB Saunders, Philadelphia, 1980.

Getty R. Sisson and Grossman's The Anatomy of the Domestic Animals. 5th ed. WB Saunders, Philadelphia, 1975.

Gilbert BM. Mammalian Osteology. 2nd ed. Missouri Archaeological Society, Columbia, MO, 1990.

Gilbert BM, Marin LD, Savage HG. Avian Osteology. Missouri Archaeological Society, Columbia, MO, 1981.

Halstead P, Collins P. Sorting the sheep from the goats: Morphological distinctions between the mandibles and mandibular teeth of adult *Ovis* and *Capra*. J Archaeol Sci 2002;29:545–553.

Milne C, Crabtree PJ. Prostitutes, a Rabbi, and a Carpenter–Dinner at Five Points in the 1830s. Hist Archaeol 2001;35(3):31–48.

Milne C, Crabtree PJ. Revealing Meals: Ethnicity, Economic Status and Diet at Five Points, 1800–1860. In: Yamin R, ed. Tales of the Five Points: Working Class Life in Nineteenth Century New York, Volume II: An Interpretive Approach to Working Class Life. General Services Administration, New York, 2002, pp. 130–196.

Mulhern DM, Ubelaker DH. Differences in osteon banding between human and nonhuman bone. J Forensic Sci 2001;46(2):220–222.

Noël Hume I. Martin's Hundred: The Discovery of a Lost Colonial Virginia Settlement. Dell, New York, 1979.

Olsen SJ. Mammal Remains from Archaeological Sites. Papers of the Peabody Museum of Archaeology and Ethnology, Harvard University 56-1. Peabody Museum, Cambridge, MA, 1964.

Olsen SJ. Fish, Amphibian and Reptile Remains from Archaeological Sites. Papers of the Peabody Museum of Archaeology and Ethnology, Harvard University 56–2. Peabody Museum, Cambridge, MA, 1968.

Owsley DW, Mann RW. Medicolegal case involving a bear paw. J Am Podiatr Med Assoc 1990;80(11):623–625.

Payne S. Morphological Distinctions between the Mandibular Teeth of Sheep, *Ovis*, and Goats, *Capra*. J Archaeol Sci 1985;12:139–147.

Raftery B. Pagan Celtic Ireland: The Emigma of the Irish Iron Age. Thames and Hudson, London, 1994.

Scheuer L, Black S. Developmental Juvenile Osteology. Academic Press, San Diego, 2000.

Schmid E. Atlas of Animal Bones for Prehistorians, Archaeologists, and Quaternary Geologists. Elsevier, New York, 1972.

Semaw S. The World's Oldest Stone Artifacts from Gona, Ethiopia: Their Implications for Understanding Stone Technology and Patterns of Human Evolution between 2.6-1.5 Million Years Ago. J Archaeol Sci 2000;27:1197–1214.

Semaw S, Rogers MJ, Quade J, et al. 2.6-Million-Year-Old Stone Tools and Associatede Bones from OGS-6 and OGS-7, Gona, Afar, Ethiopia. J Human Evol 2003;45:169–177.

Silver IA. The ageing of domestic animals. In: Brothwell D, Higgs E, eds. Science in Archaeology, Praeger Publishers, New York, 1969, pp. 283–302.

Solecki RS. The First Flower People. Alfred Knopf, New York, 1971.

Steele DG, Bramblett CA. The Anatomy and Biology of the Human Skeleton. Texas A&M University Press, College Station, TX, 1988.

Stewart T. Essentials of Forensic Anthropology. Charles C. Thomas, Springfield, IL, 1979.

Symes SA, Berryman HE, Smith OC. Saw marks in bone: Introduction and examination of residual kerf contour. In: Reichs KJ. Forensic Osteology, 2nd ed. Charles C Thomas, Springfield, IL, 1998.

Symes SA, Williams JA, Murray EA, et al. Taphonomic context of sharp-force trauma in suspected cases of human mutilation and dismemberment. In: Haglund WD, Sorg MH, eds. Advances in Forensic Taphonomy, CRC Press, Boca Raton, FL, 2002, pp. 403–434.

Ubelaker DH. Human Skeletal Remains: Excavation, Analysis, Interpretation. Taraxacum, Washington, 1989.

Wailes B. Dún Ailinne: A Summary Excavation Report. Emania 1990;7:10–21.

Wailes B. Irish Royal Sites. In Ancient Europe 8000 B.C.–A.D. 1000. In: Bogucki P, Crabtree P. Encyclopedia of the Barbarian World, Vol. 2., Scribner's, New York, 2004, pp. 239, 240.

West SE. West Stow: the Anglo-Saxon Village. East Anglian Archaeology, No. 24. Suffolk County Planning Department, Ipswich, UK, 1985.

White TD. Human Osteology. 2nd ed. Academic Press, San Diego, 2000.

White TD, Folkens PA. Human Bone Manual. Elsevier Academic Press, San Diego, 2005.